Praise for

THE PROCRASTINATION EQUATION

"Procrastinating just makes unpleasant tasks worse, so why is it so hard for us to resist dithering and delay? *The Procrastination Equation* is crammed with surprising insights about procrastination and human nature—as well as concrete, helpful solutions for fighting procrastination."

—Gretchen Rubin, author of *The Happiness Project*

"*The Procrastination Equation* is this season's must-read self-help book. In addition to offering useful strategies to fight a common problem, it's a fascinating read." —*Montreal Gazette*

"An upbeat, motivational guide to procrastination. . . . Everything you ever wanted to know about procrastination but never got around to reading." —*Kirkus Reviews*

"The book itself is great. Even the argument behind that sentence is persuasive." —*The Guardian*

"In his absorbing first book Steel looks closely at the often misunderstood habit of procrastination. His engaging guide will appeal to a wide audience of past, present, and future procrastinators and researchers trying to get a handle on the science of putting things off."

—*Publishers Weekly*

"Procrastination is the saffron spice of human behavior, where even small amounts of this tendency can shatter the best of intentions. In this illuminating book Piers Steel shows us the secrets of procrastination, how it affects us, and how we will, one day, be able to prevail."

—Dan Ariely, author of *The Upside of Irrationality* and *Predictably Irrational*

"Steel takes us through the history of procrastination, showing how it has become, in modern times, a serious problem that leads to increased health troubles, loss of productivity, and unnecessary poverty or depression. Fortunately, he also suggests ways we can stop procrastinating and get ourselves on track. A useful, eye-opening book. Now, if only the people who most need to read it could find the time to do so."

—*Booklist*

"Why you 'put off till tomorrow what you can do today' forms the crux of Steel's book, in which he not only answers that question but details specific techniques to reign in the impulse. . . . Offers good advice."

—*Library Journal*

"I put off writing this blurb until the last minute. I thought it was because I was too busy, but after reading *The Procrastination Equation*, I know the real reasons. Piers Steel will help you tackle the goals—big and small—that always seemed just a little out of reach."

— Richard Florida, author of *The Great Reset*

THE PROCRASTINATION EQUATION

How to Stop Putting Things Off and Start Getting Stuff Done

PIERS STEEL, PhD

HARPER

NEW YORK • LONDON • TORONTO • SYDNEY

HARPER

A hardcover edition of this book was published in 2011 by HarperCollins Publishers.

HarperCollins books may be purchased for educational, business, or sales promotional use. For information please write: Special Markets Department, HarperCollins Publishers, 10 East 53rd Street, New York, NY 10022.

Published simultaneously in Canada in 2010 by Random House Canada, a division of Random House of Canada Limited, Toronto.

FIRST HARPER PAPERBACK PUBLISHED 2012.

Library of Congress Cataloging-in-Publication Data is available upon request.

ISBN 978-0-06-170362-1 (pbk.)

13 14 15 16 OFF/RRD 10 9 8 7 6 5 4 3

"To my brother Toby. He knew that the clock is always ticking."

CONTENTS

AUTHOR'S NOTE

Procrastination has been my life's work—both as a researcher and as a practitioner. With research so often being "me-search," this isn't accidental. Scientists often intimately know the subjects they study—they are problems they themselves face. It's true that I have sympathy for the procrastinator's plight because it is one I shared for many years.* Nowadays my work has received international acclaim, I have coached national college champions in business school competitions, and awards for teaching and research hang on my office wall. But for most of my life, I felt potential languishing inside me mingled with frustration because I couldn't sustain any of my many attempts to improve. Encountering people who were naturally more capable of getting things done simply reminded me

* And everyone knew it. Here is an excerpt from a letter my late brother sent to my uncle: "Have you heard from Piers about his research? He has forged himself into an expert on procrastination, publishing numerous articles on the subject and being interviewed on national radio and in the press. I get a chuckle, as Piers was the worst procrastinator during his high school and college years."

of my own deficiencies, curdled my spirit, and raised considerable misplaced resentment. Luckily, I was attracted to a profession whose very purpose was to identify the key enablers of change, which I then systematically put into practice in my own life one by one.

My PhD is in Industrial/Organizational Psychology, the scientific study of our actions and minds in the workplace. Psychology applied to work focuses on how to improve people's performance, well-being, and, quite appropriately, motivation or lack thereof. Unfortunately, many of the techniques of this discipline aren't well known, buried in the depths of obscure journals and written in scholarly language comprehensible only to the initiated. For procrastination, the problem gets even more complex. This subject has attracted the attention of all the social sciences and inspired research around the world. With over eight hundred scientific articles on the topic from fields spanning economics to neuroscience, in languages ranging from German to Chinese, the challenge is to find and make sense of them all.[1] And this is where I come in. I found two ways to study procrastination. The first was by doing my own research, which you will read about presently. That gave me the basis for a theory of how and why we put things off. But then I needed to deal with the panoply of disciplines that have studied procrastination and published results in so many different journals and books. I was lucky enough to stumble upon meta-analysis, a recently developed scientific technique, and adapt it for my research.

Meta-analysis mathematically distills the results from thousands of studies to their core consensus. At a basic level, meta-analysis is what lets science progress. By enabling a synthesis of knowledge, it reveals the underlying truths we seek. It is very powerful, it has applications in every field, and it increasingly provides the information we need to run the world. The medical treatment you get from your doctor, for example, is likely based on the results of meta-analyses, from asthma to Alzheimer's.[2] It is a discipline I have mastered: I have created some of its basic techniques, I teach it to others, and I have developed software for it. I like to think of it as something I am good at.[3] It was natural, then, to meta-analyze the body of research on procrastination, given that there was no other way to put together all the findings. I have to say that the field of procrastination proved to be daunting, as almost every possible scientific methodology and technique has been thrown at it. Researchers have run laboratory experiments, read through personal diaries, twiddled with neurotransmitters, and dissected DNA. They have monitored every setting, from airports to shopping malls; they have wired entire classrooms to track every student's twitch and shudder; and they have studied procrastinators from every background, including pigeons, vermin, and members of the U.S. Congress. Making them all fit coherently together was like being a conductor of a madhouse orchestra. The strings, woodwinds, brass, and percussion are all playing the same tune but not in the same room, in the same rhythm, or in the same key. Turning that noise into music is what this book is about.

What I found will surprise you and challenge the status quo. Some of my work has already been published, such as my article "The Nature of Procrastination," which appeared in *Psychological Bulletin,* the social sciences' most respected journal. Some of it has already been reported in hundreds of media venues around the world, from India to Ireland and from *Scientific American* to *Good Housekeeping* and *The Wall Street Journal.* But most of what I found is presented here for the first time. Within these pages, you will find out that we've been misdiagnosing procrastination for decades, attributing it to a trait associated with less procrastination, not more. The real reasons for procrastination are partly genetic and can be traced to the fundamental structure of our brains, which is why procrastination is seen in every culture and throughout history. The environment, however, isn't blameless; it may not be responsible for procrastination's existence, but it is responsible for its intensity—modern life has elevated procrastination into a pandemic. And guess what? All these findings follow from the application of a simple mathematical formula I devised—the Procrastination Equation.

Because I was able to tease out the fundamentals of the dynamic that makes us procrastinate, I have also been able to figure out strategies that we can use throughout our lives—school, work, or personal—to combat our innate tendency to put things off. A tall order? You bet. That's why it has taken me so many years to write this book. I hope the hours you spend reading it will reward you with a new way of thinking about how to spend—and waste—your time.

CHAPTER ONE

Portrait of a Procrastinator

Never put off till tomorrow,
what you can do the day after tomorrow.
MARK TWAIN

This book is about every promise you made to yourself but broke. It is about every goal you set but let slide, never finding the motivation. It is about diets postponed, late-night scrambles to finish projects, and disappointed looks from the people who depend on you—or from the one you see in the mirror. It is about being the slacker in your family and the straggler in your circle of friends. It is about that menacing cloud of uncompleted chores, from the late bill payments to the clutter that fills your home. It is about that doctor's appointment you have been putting off and the finances still in disarray. It is about dawdling, delay, opportunity lost, and more. Much more. This book is also about the *other* side, the moments of action when procrastination gives way to crystal clarity and attention, work is devoured without hesitation, and giving up never even occurs to you. It is about personal transformation, about unencumbered desire free of internal competition, and the guiltless leisure you can enjoy when your daily tasks are done. This book is about potential, wasted and fulfilled; about dreams that fade into obscurity and dreams we can make come true. Best of all, this book is about shifting the rest of your life away from putting it off to getting it done.

The pivot point that tips us away from accomplishing what we want and need to do is procrastination. It isn't a question of laziness, although the two are easily confused. Unlike the truly slothful, procrastinators want to do what they need to do—and usually do get around to it, but not without a lot of struggle. I will show that this dillydallying is in part hereditary, and that we are hardwired to delay. Our tendency to put things off took a hundred million years to form and is now almost etched into our being. But research shows that, despite its ingrained nature, we can modify our habits and change this behavior. Procrastinators who understand the processes behind their inaction can master them and become less stressed about their deadlines and more able to meet them.

This book tells procrastination's story. It stretches from Memphis of ancient Egypt to modern New York City, from the cancer ward to the stock market floor. I hope to enlighten you about why we procrastinate, what comes of procrastination, and what strategies we can employ to do something about it. We will start off simply, establishing what procrastination is, helping you decide whether you are a procrastinator, and if so, how you likely experience a bout of procrastination. If you are a procrastinator—and the odds are good that you are—you are part of a very large community indeed. It is time we all got to know each other a little bit better.

WHAT PROCRASTINATION IS AND ISN'T

There is so much confusion about procrastination that it is best to lay our subject bare on the dissecting table and start immediately separating the dilly from the dally. By procrastinating you are not just delaying, though delay is an integral part of what you are doing. Procrastination comes from the Latin *pro,* which means "forward, forth, or in favor of," and *crastinus,* which means "of tomorrow." But procrastination means so much more than its literal meaning. Prudence, patience, and prioritizing all have elements of delay, yet none means the same as procrastination. Since its first appearance in the English language in the sixteenth century, procrastination has identified not just any delay but an *irrational* one—that is, when we voluntarily put off tasks despite believing ourselves to be worse off for doing so. When we procrastinate, we know we are acting against our own best interests.

Still, you will find people mischaracterizing wise delays as procrastination. Seeing a co-worker stretched out in his office chair, arms crossed behind his head, relaxed, you ask what he is up to and get a cheerful response of "Me? I'm procrastinating!" But he isn't. He is happily putting off a report because he knows there is a good chance that the project is going to be cancelled later this week, and if it isn't, well, he can still definitely write it at the last minute anyway. This is smart. In this scenario, it is the person who compulsively has to finish everything as soon as possible who is irrational, tackling work even when it is destined to become irrelevant. The obsessive who completes every task at the first opportunity can be just as dysfunctional as the procrastinator who leaves everything to the last moment. Neither one is scheduling time intelligently.

Consequently, it isn't procrastination if you fail to arrive at a party far earlier than everyone else or if you don't get to the airport for your flight three hours in advance. By delaying a little bit, you save awkward moments with your host, who is likely still getting things ready, and you will be spared uncomfortable hours at your gate waiting for your plane to take off. Neither is it procrastination to respond to emergencies by dropping (and putting off) everything else. Insisting that you should finish mowing the front lawn before attending to your house, which has just caught fire, isn't smart. Sure, you didn't put off trimming the grass, but the charred ruin of your home is too high a price to pay. Alternatively, flexibly adapting your schedule to respond to the pressing needs of a spouse or a child will likely save you from ruining your family. Not everything can happen at once; it is in your choice of what to do now and what to delay that procrastination happens, not in delay itself.

YOU THE PROCRASTINATOR

Now that we understand what procrastination is, do you practice it? Where do you land in the ranks of procrastination? Are you a garden-variety dillydallier or are you hardcore with "tomorrow" tattooed across your back? There are some entertaining methods that may reveal your propensity to procrastinate. To begin, check your handwriting. If it is sluggish and disjointed, it may indicate you are likewise. Alternatively, look to the stars . . . well, really the planets. Astrologers note that when Mercury is in retrograde or in opposition to Jupiter, procrastination tends to be on the uptick.[1] Or try a tarot card reading. The "Two of Swords" often indicates you are split with a dilemma and procrastinating on your decision. Personally, I prefer a more scientific approach.

You can go to my website, www.procrastinationequation.com, for a comprehensive test that I've administered to tens of thousands of subjects, and compare your level of irrational delay with those of individuals around the world. However, if time is pressing and you wish not to delay, you might try the shorter quiz provided below. Complete the mini-version here by circling your response to each of these nine items and then calculating the total. Note that questions 2, 5, and 8 are scored in the opposite direction from the other items:

		VERY SELDOM OR NOT TRUE OF ME	SELDOM TRUE OF ME	SOMETIMES TRUE OF ME	OFTEN TRUE OF ME	VERY OFTEN TRUE OR TRUE OF ME
1.	I delay tasks beyond what is reasonable.	1	2	3	4	5
2.	I do everything when I believe it needs to be done.	5	4	3	2	1
3.	I often regret not getting to tasks sooner.	1	2	3	4	5
4.	There are aspects of my life that I put off, though I know I shouldn't.	1	2	3	4	5
5.	If there is something I should do, I get to it before attending to lesser tasks.	5	4	3	2	1
6.	I put things off so long that my well-being or efficiency unnecessarily suffers.	1	2	3	4	5
7.	At the end of the day, I know I could have spent the time better.	1	2	3	4	5
8.	I spend my time wisely.	5	4	3	2	1
9.	When I should be doing one thing, I will do another.	1	2	3	4	5

TOTAL SCORE ________

SCORE	COMPARED TO EVERYONE ELSE	
19 or less	You are in the bottom 10%	Your mantra is "first-things-first"
20-23	You are in the bottom 10-25%	
24-31	You are in the middle 50%	Average procrastinator
32-36	You are in the top 10-25%	
37 or more	You are in the top 10%	Tomorrow is your middle name

Where did you end up? Are you legendary for leaving things to the last minute or do you only put off exercising and taxes, like almost everyone else?

PROCRASTINATION POLKA

The higher you scored on that procrastination test, the greater the chance that you are procrastinating right now. Certain other tasks should be occupying your attention—which sadly means you have better things to do than reading this book. These tasks are likely unpleasant, possibly administrative and boring, and perhaps difficult to visualize as being successfully accomplished. Let me make a few guesses about what is on your plate:

- Is your laundry basket overflowing?
- Are there dirty dishes in the sink?
- Do your smoke detectors need new batteries?
- How about your car battery? What is the air pressure in your tires and how long has it been since the last oil change?
- Isn't there a ticket to book, a room to reserve, a bag to pack, a passport to renew?
- Have you informed your boss about your vacation plans?
- Have you bought a gift for that upcoming birthday?
- Have you filled out your time sheets, performance reviews, and expense reports?
- Did you hold that difficult conversation with the employee whose work is not up to par?

- Have you scheduled the meeting you are dreading?
- What about the big project your boss gave you? Are you making progress?
- Did you make it to the gym this week?
- Have you called your mom?

How does that list strike you? You can add to it, of course. Even if I didn't score a direct hit, you were likely procrastinating somewhere else, pushing a task into the future. On its own, each of these postponed tasks has few repercussions. Together, they can culminate in misery by nibbling away at your life. The major project, the one with the hard deadline, is the mother of all such concerns; it can keep you awake at night and make it difficult to accomplish any of the other tasks on your list. At one time or another, we have *all* felt motivationally marooned and unable to get around to the report, the research, the writing, the presentation to prep, or the exam to ace.

There is a common pattern to all procrastination and it goes something like this. At the start of a big project, time is abundant. You wallow in its elastic embrace. You make a few passes at getting down to it, but nothing makes you feel wholeheartedly engaged. If the job can be forgotten, you'll forget it. Then the day arrives when you really intend to get down to work; but suddenly it's just something you don't feel like doing. You can't get traction. Every time you try to wrap your mind around it, something distracts you, defeating your attempts at progress. So you forward your task to a date with more hours, only to find that every tomorrow seems to have the same twenty-four. At the end of each of these days, you face the disquieting mystery of where it went. This goes on for a while.

Eventually, time's limited nature reveals itself. Hours, once tossed carelessly away, become increasingly limited and precious. That very pressure makes it hard to get started. You want to get going on the big project but instead you take on peripheral chores. You clean your office or clean up your e-mail; you exercise; you shop and cook. Part of you knows this isn't what you should be doing, and so you say to yourself, "I am doing this; at least I am preparing by doing something." Eventually, it is too late in the day to really get started, so you may as well go to bed. And the cycle of avoidance starts again with the dawn.

Sometimes, to quell your anxiety, you give in to total diversion. You take a moment to check your e-mail or the sports scores. From there, why not respond to a few messages or watch a few minutes of TV? Soon these temptations have seduced you. The task still waggles itself in the periphery of your vision, but you don't want to look it in the eye—it will have you if you look—so you burrow deeper into your distractions. You write long passionate comments on online forums, troll for news tidbits, or manically switch TV channels at the first ebb of interest. Pleasure turns to powerlessness as you become unable to extract yourself.

As the deadline approaches, you make the diversions more intense so that they will sufficiently distract you. Banishing anything that reminds you of the dreaded thing, you shun calendars and timepieces. In a willful distortion of reality, you shift your plans from what you once could solidly accomplish to what is minimally possible. When you should be working harder than ever, you are sleeping in, daydreaming of alternative worlds, of winning the lottery, of being anywhere but here. As anxiety mounts, you want immediate relief, escape, rewards—anything that gives you the illusion of safe harbor. If friends or relatives or co-workers try to separate you from your diversions, you meet them with an annoyed: "Just a minute! I'LL DO IT AFTER THIS!" Unfortunately, "this" never ends. Secretly, you are full of self-recrimination and self-doubt, envious of those who simply get things done.

Energy builds until finally a threshold is crossed and something clicks. You start working. Some inner mind has quietly boiled the task down to its essence, as there are no more moments to spare. You wade into the work, making ruthless decisions and astonishing progress. In place of that menacing cloudiness, a glittering clarity comes over you. There is purity to your work, fueled by the real urgency of now or never. For a lucky few, this surge of efficiency will enable them to get the project done. For others, this initial rush wanes before the cursed thing is completed. After too many hours of sleepless concentration, brains shut down. Caffeine and sugar only offer an unsatisfying buzz. Tick, tock . . . the time has run out. You limp across the finish line with insufficient preparation, giving the world your second best.

This is so common as to be unremarkable—except to the person who has suffered through the experience and knows the performance was not up to par. The relief at getting a job done doesn't always make up for doing a sloppy job. Even if you managed to perform brilliantly, the achievement is tainted with a whiff of what might have been. And this kind of procrastination has likely cast a cloud on an evening out, a party, or a vacation, which you couldn't fully enjoy because half of your mind was elsewhere, obsessing about what you were avoiding. You resolve that this will never happen again; the cost of procrastination is too great.

The trouble with such resolutions is that procrastination is a habit that tends to endure. Instead of dealing with our delays, we excuse ourselves from them—self-deception and procrastination often go hand-in-hand.[2] Exploiting the thin line between *couldn't* and *wouldn't,* we exaggerate the difficulties we faced and come up with justifications: a bad chest cold, an allergic reaction that caused sleepiness, a friend's crisis that demanded our attention. Or we deflect responsibility entirely by saying, "Gee whiz, who knew?" If you couldn't have anticipated the situation, then you can't be blamed. For example, how would you respond to the following questions regarding your last bout of procrastination?

- Did you know the task was going to take so long?
- Did you realize that the consequences of being late were so dire?
- Could you have expected that last-minute emergency?

The honest answers are likely yes, yup, and definitely, but it's difficult to answer honestly, isn't it? And that is the problem.

Some procrastinators will even try to frame their self-destructive inaction as a thoughtful choice. For example, is it wrong to put off your career to pursue more family time? It depends on who you are. Some people relish the work-focused model of success, resenting time taken away from the job, and so they may miss out on family dinners and school plays. Others prosper in the home and community, enjoying the relationships nurtured there, at the expense of tasks at work. To the casual observer, it isn't easy to tell which choice is procrastination and which is a purposeful decision. Only the procrastinator knows for sure.

In the back of their minds, many procrastinators hope they won't need excuses. They bank on Lady Luck. Sometimes it works. Frank Lloyd Wright drew his architectural masterpiece, Fallingwater, in the three hours before his patron, Edgar Kaufmann, came to see the sketches. Tom Wolfe cranked out in a midnight panic forty-nine pages of almost unedited prose for an *Esquire* magazine piece on California's hotrod and custom car culture. Byron Dobell, his editor, simply removed "Dear Byron" from the top of Wolfe's memo and printed it under the title "There Goes (Varoom! Varoom!) That Kandy-Kolored Tangerine-Flake Streamline Baby," and a new style of journalism was born. But I don't need to tell you how rare such outcomes are. By your own standards, if you thought delay was a good idea in the first place, you wouldn't be procrastinating.

THE PROCRASTINATOR'S PROFILE

If it makes you feel any better, procrastination puts us in good company. It's as common as morning coffee. Across scores of surveys, about 95 percent of people admit to procrastinating, with about a quarter of these indicating that it is a chronic, defining characteristic.[3] "To stop procrastinating" is at any time among the world's top reported goals.[4] Procrastination is so prevalent that it has its own brand of humor. Possibly the best excuse for missing a deadline came from Dorothy Parker. When asked by *The New Yorker*'s editor, Harold Ross, for a piece that was late, she woefully explained, using her dark and sorrowful eyes to full effect, "Somebody was using the pencil." And, of course, there is the most infamous of all procrastination jokes. Don't you know it? I will tell you later.

No occupational category seems immune from procrastination, but writers seem especially prone. Agatha Christie was guilty of it and Margaret Atwood admitted she often spends "the morning procrastinating and worrying, and then plunges into the manuscript in a frenzy of anxiety around 3:00 p.m." Newscasters can also suffer from it; witness Ted Koppel's quip: "My parents and teachers used to be exasperated by the fact I would wait until the last minute, and now people are fascinated by it."[5] Procrastinators come from every letter of the occupational alphabet, from astronauts to Episcopalian priests and from X-ray technicians to zookeepers.[6] Unfortunately, whatever the job, procrastinators are more likely to be unemployed or working part-time compared to their non-procrastinating counterparts. Procrastinators can be of either sex, though the Y chromosome has a slight edge. A group of a hundred hardened procrastinators would likely be composed of 54 men and 46 women, leaving 8 unmatched males vying for a female dalliance. You see, procrastinators tend to be available . . . sort of. They are more likely to be single than married but also more likely to be separated than divorced. They put off ending as well as beginning commitment. Age also determines procrastination.[7] As we progress from grade school through to the retirement home and the closer we come to life's final deadline, the less we put off. Those who have matured physically are, unsurprisingly, more mature in character.

This demographic exploration, though interesting, isn't as useful as identifying procrastinators by their psychological profile. There is indeed a core trait explaining why we put off, but it might not be what you have heard. It is commonly thought that we delay because we are perfectionists, anxious about living up to sky-high standards.[8] This perfectionist theory of procrastination sounds good and even feels good. Perfectionism can be a desirable trait, as shown by the canned response to the interview question, "What is your biggest weakness?" When Bill Rancic was asked that question just before winning the first season of Donald Trump's *The Apprentice,* he replied, "I'm too much of a perfectionist; it's a flaw," prompting his interviewer to interject, "Being a perfectionist is a good thing; it means you keep striving." But the perfectionism-procrastination theory doesn't pan out. Based on tens of thousands of participants—it's actually the best-researched topic in the entire procrastination field—perfectionism produces a negligible amount of procrastination. When the counseling psychologist Robert Slaney developed the Almost Perfect Scale to measure perfectionism, he found that "perfectionists were *less* likely to procrastinate than non-perfectionists, a result that contradicted the anecdotal literature."[9] My research backs him up: neat, orderly, and efficient perfectionists don't tend to dillydally.[10]

How, then, did we come to believe that perfectionism causes procrastination? Here is what happened. Perfectionists who procrastinate are more likely to seek help from therapists, so of course they turn up in clinical research about procrastination in greater numbers. Non-perfectionist procrastinators (and for that matter, non-procrastinating perfectionists) are less likely to seek professional help. Perfectionists are more motivated to do something about their failings because they are more likely to feel worse about whatever they are putting off. Consequently, it is not perfectionism that is the problem but the *discrepancy* between perfectionist standards and performance.[11] If you are a perfectionist and are suffering from high standards that are unachievable, you might want to do something about that too, but you will need an additional book: this one is about procrastination.

What is really the main source of procrastination? Thirty years of research and hundreds of studies have isolated several personality traits that predict procrastination, but one trait stands above the rest. The Achilles Heel of procrastination turns out to be *impulsiveness;* that is, living impatiently in the moment and wanting it all now.[12] Showing self-control or delaying gratification is difficult for those of us who are impulsive. We just don't have much ability to endure short-term pain for long-term gain.[13] Impulsiveness also determines how we respond to task anxiety. For those of us who are less impulsive, anxiety is often an internal cue that gets us to start a project early, but for those who are more impulsive it is a different story: anxiety over a deadline will lead straight to procrastination.[14] The impulsive try to avoid an anxiety-provoking task temporarily or block it from their awareness, a tactic that makes perfect sense if you're thinking short term. In addition, impulsiveness leads procrastinators to be disorganized and distractible or, as my colleague Henri Schouwenburg puts it, to suffer from "weak impulse control, lack of persistence, lack of work discipline, lack of time management skill, and the inability to work methodically."[15] In other words, impulsive people find it difficult to plan work ahead of time and even after they start, they are easily distracted. Procrastination inevitably follows.

LOOKING FORWARD

So there it is. Procrastination is pervasive. Almost as common as gravity and with an equal downward pull, it is with us from the overfull kitchen garbage can in the morning to the nearly empty tube of toothpaste at night. In the next chapter, I'll let you in on the research that has helped me understand why we delay things irrationally and why procrastination is so widespread. I'll reveal and explain the Procrastination Equation, a formula that shows the dynamics of this way of behaving, and then I'll tell you about the amazing opportunity I had to study this phenomenon in the real world. Subsequent chapters will describe the different elements that are at play in our minds and hearts, and then we'll look at the price of procrastination in our lives and in society at large. There's always a good side to the kind of research I present—within the causes we can also find the cures. So the last part of the book will offer ways in which individuals, bosses, teachers, and parents can improve their own motivation and motivate others, in the hope that procrastination will be less of a scourge. The final chapter pushes you to put these proven practices into your own life. The advice here is evidence-based, as scientifically vetted and pharmaceutically pure as it gets; it's the good stuff from behind the counter, so don't overdo it.

CHAPTER TWO

The Procrastination Equation

THE RESULT OF EIGHT HUNDRED STUDIES PLUS ONE

My own behavior baffles me. For I find myself doing what I really hate, and not doing what I really want to do!

ST. PAUL

Rejection is wearing thin on Eddie during his first sales job. He attentively attended each sales seminar, read all the recommended books, and dutifully repeats the positive affirmations "I can do it! I am a winner!" each morning in the mirror. Still, after another day without a sale he is looking at his phone with dread. As he picks it up and cold calls another prospect, the only response he anticipates hearing is yet another "I am in a meeting" or "click" as they hang up halfway through his introduction. Indeed, he is brushed off once again. "What is the point?" Eddie asks himself. Demoralized, he organizes his desk, fills out all the paperwork to update his benefits package, and surfs the Internet to get insights about competitors' products. He puts off his phone calls until later—the dregs of the day when most of his potential clients are leaving for home. His boss checks in on him and recognizes the signs. Eddie's decision to delay is the beginning of the end of his sales career.

Valerie's face is as blank as her computer screen. She stares at it, knowing that words should be there, words written by her, but nothing appears. Not even a letter. "Why? Why?" she wonders. It is not like she hasn't done pieces like this before, but for some reason this assignment on municipal politics due tomorrow is mind numbing. "Write," she thinks, "Press your fingers into the keyboard." In response, "asdfkh" appears on the screen. Better than nothing. Convincing herself she needs a short break from interminable boredom, Valerie starts texting her friends who direct her to a nifty website spoofing popular bands. After watching a few music videos, she finds a satire site on television shows and texts that link back to her friends. Soon, Val's virtual group is trying to one-up each other to find the funnier and cleverer clip. Hours go by, then it dawns on her that it is near the end of day and she feels even less inspired than when she first chose to take that "short break." She dives into the writing but the end product reflects the effort and time put into it. It's crap.

The vacation plans are set! Tom is for once ahead of the ball, booking time off in advance to fly to the Dominican Republic. Thanks to his foresight he even paid for his flight on points from his frequent-flyer program. The only detail left is to reserve a room at a hotel, but that can be done at anytime. But what can be done at anytime is often done at no time. As the months slip by, Tom pushes the task forward to each subsequent week or forgets about it altogether. There is always something more pressing to attend to, like his favorite television show. Finally, as he thinks about what to pack, he realizes that there are no more weeks to push the task forward into and that he has left it far too late. He goes online and, finding little available, makes a hurried and haphazard reservation. When his plane later sets down in the Dominican, he hopes that his hotel is as beautiful as the island. It isn't. It's too far from the beach, his room is decorated with dead mosquitoes, adjoins a disgusting bathroom, and the hotel dining gives him food poisoning.

Eddie, Valerie, and Tom are all procrastinators but they are not identical. Just as a car can stop running because of an empty gas tank, a blown tire, or a dead battery, there are a multitude of causes for procrastination—even if the outward behavior is the same. Eddie, Valerie, and Tom all procrastinated for different underlying reasons and each one represents a facet of the Procrastination Equation, the mathematical formula I derived that describes irrational delay. Understanding why Eddie, Valerie, and Tom put off their respective tasks is the essence of this book. To this end, we are going to do a little more assessment. In the last chapter, we established the degree to which you procrastinate. In this chapter, we are going to find out why you put things off. Are you an Eddie, a Valerie, a Tom or some hybrid of all three? Take this test by circling your response to each of the following 24 items and find out:

VERY SELDOM OR NOT TRUE OF ME	SELDOM TRUE OF ME	SOMETIMES TRUE OF ME	OFTEN TRUE OF ME	VERY OFTEN TRUE OR TRUE OF ME

1. When I put in the hours, I am successful.

   1 2 3 4 5

2. Uninteresting work defeats me.

   1 2 3 4 5

3. I get into jams because I will get entranced by some temporarily delightful activity.

   1 2 3 4 5

4. When I apply myself, I see the results.

   1 2 3 4 5

5. I wish my job was enjoyable.

   1 2 3 4 5

6. I take on new tasks that seem fun at first without thinking through the repercussions.

   1 2 3 4 5

7. If I try hard enough, I will succeed.

   1 2 3 4 5

8. My work activities seem pointless.

   1 2 3 4 5

9. When a temptation is right before me, the craving can be intense.

   1 2 3 4 5

10. I am confident that my efforts will be rewarded.

    1 2 3 4 5

11. Work bores me.

    1 2 3 4 5

12. My actions and words satisfy my short-term pleasures rather than my long-term goals.

    1 2 3 4 5

	VERY SELDOM OR NOT TRUE OF ME	SELDOM TRUE OF ME	SOMETIMES TRUE OF ME	OFTEN TRUE OF ME	VERY OFTEN TRUE OR TRUE OF ME

13. I am persistent and resourceful.

    1 2 3 4 5

14. I lack enthusiasm to follow through with my responsibilities.

    1 2 3 4 5

15. When an attractive diversion comes my way, I am easily swayed.

    1 2 3 4 5

16. Whatever problems come my way, I will eventually rise above them.

    1 2 3 4 5

17. When a task is tedious, again and again I find myself pleasantly daydreaming rather than focusing.

    1 2 3 4 5

18. I have a hard time postponing pleasurable opportunities as they crop up.

    1 2 3 4 5

19. I can overcome difficulties with the necessary effort.

    1 2 3 4 5

20. I don't find my work enjoyable.

    1 2 3 4 5

21. I choose smaller but more immediate pleasures over those larger but more delayed.

    1 2 3 4 5

22. Winning is within my control.

    1 2 3 4 5

23. If an activity is boring, my mind slips off into other diversions.

    1 2 3 4 5

24. It takes a lot for me to delay gratification.

    1 2 3 4 5

To score, add up your answers to each of the following questions:

Eddie's Scale = 1 + 4 + 7 + 10 + 13 + 16 +19 + 22 =

Valerie's Scale = 2 + 5 + 8 + 11 + 14 + 17 + 20 + 23 =

Tom's Scale = 3 + 6 + 9 + 12 + 15 + 18 + 21 + 24 =

If you scored 24 or *lower* for *Eddie's Scale,* you have some similarities with his situation. On the other hand, if you scored 24 or *higher* for *Valerie's Scale* or *Tom's Scale,* you really should give them a call as you have a lot in common. You see, Eddie, Valerie, and Tom represent respectively the three basic elements of motivation: Expectancy, Value, and Time. Once you grasp their situations, you will understand the components of the Procrastination Equation. After this, we will look at how each of these pieces fits together with the others to form the overall formula. Yes, there will be math, but don't balk. A version of this principle was illustrated within just two glossy pages of *Yes! The Science Magazine for Kids.* If twelve-year-olds can get it, so will you.

LOW EXPECTANCY EDDIE

Eddie's story is regrettably common in sales. Rejection is part and parcel of the job, and most sales people receive an ungodly number of "no's" before they get a "yes," especially at the beginning of their careers. Many aspiring salespeople, like Eddie, succumb to this steady stream of rebuffs and find themselves lacking the motivation to perform; it takes especially resilient people to rise above relentless negativity. What is sapping Eddie's motivation and causing his procrastination? It is *Expectancy*—what he *expects* will happen. After a series of attempts that all resulted in failure, he began to expect failure even before he started. High expectancy forms the core of self-confidence and optimism: but if you start believing your goals aren't achievable, you stop effectively pursuing them.

Consequently, if during your self-assessment you *disagreed* with statements like "I am confident that my efforts will be rewarded" or "Winning is within my control," you are like low-expectancy Eddie.

The results from thirty-nine procrastination studies consisting of almost seven thousand people indicate that while some procrastination stems from overconfidence, the opposite is far more common. Procrastinators are typically less confident, especially about the tasks they are putting off. If you are procrastinating about schoolwork, you likely consider the assignment difficult. If you are procrastinating about getting healthier, by starting an exercise program, for example, or by eating better, odds are that you question your ability to follow through. And, if you are unemployed, you are likely procrastinating on your job search because you are discouraged about your chances of getting hired.

The seminal work of Martin Seligman, one of the leaders of the positive psychology movement, demonstrates the connection between lack of self-confidence or optimism and procrastination.[1] If you love dogs, as I do, please try to forgive Dr. Seligman; he experimented by jolting canines with electricity.* The gist of what he did was to yoke together two sets of dogs, and zap them at random intervals. Both sets received the same electric shocks and for the same duration, but the first group could press a lever that terminated the shocks for all the dogs. The second group had no control and was entirely dependent on their counterparts for ending the agony. Seligman then changed the setup; he tested both sets of dogs again but this time in a shuttle-box divided by a low barrier. One side of the box became electrified and *all* the

* An ethical choice that Dr. Seligman struggled with, as he recounts in his book *Learned Optimism*. He discontinued this experimental method as soon as he had obtained the data he needed.

dogs had the possibility of escaping simply by jumping over the partition. The first group of dogs, who previously had control of the lever, readily learned to jump over the barrier. The second group had also learned something from their previous experience. When the box was electrified, they didn't jump, but lay down and took the shock. Like low-expectancy Eddie, these dogs had learned that what they did made no difference; they had learned they were helpless.[2]

Learned helplessness is connected to quickly giving up, whether in a complacent acceptance of a prolonged sickness or in a lackluster school performance. Learned helplessness also helps explain why putting off decisions more than usual is one of the symptoms of depression.[3] The underlying cause is reduced self-confidence, which makes it difficult to invest in any demanding work.[4] On balance, a degree of learned helplessness is common. Many of us have been in situations where our world was seemingly not set up for our success. For low-expectancy Eddie, it was his sales job; for someone else it may have been a harsh upbringing in which family or classmates enforced rigid roles. Restraining beliefs can become internalized and be carried within us long after we leave the home or schoolyard where they started. Our learned self-perception becomes a self-fulfilling prophecy—by expecting to fail, we make failure a certainty. We never dig in and really try, and the end result is more procrastination.

VALERIE WITHOUT VALUE

How do you feel about what you are putting off right now? As you reflect on the question, you will be channeling Valerie. Like her, with her sluggish attempts at writing on municipal politics, we all tend to put off whatever we dislike. Consequently, that chore you are currently deferring is probably something you don't especially enjoy. The technical term for this measure of enjoyment is *Value* and the less of it a task has for you, the harder it is for you to get started on it. We have no problem initiating lengthy conversations with friends over a few drinks and a decadent dessert, but most of us delay starting on our taxes or cleaning out our basement. Similarly, the top reason that students give for essay procrastination is that they "really dislike writing term papers."[4]* Although the fact that we are less likely to promptly pursue an unpleasant task may seem pretty obvious, the scientific field lacks your insight. Scientists have committed over a dozen studies involving a good two thousand respondents to reach the same conclusion. Well, at least now we are sure.

To the degree that some tasks are universally burdensome, they reveal some touchstones of procrastination.[5] Since everyone wants to put off whatever they detest, it is no surprise that we commonly avoid cleaning up, or organizing, or seeing our doctor or dentist.[6] Since many find exercising an imposition, 70 percent of us rarely use our long-term gym memberships.[7] Similarly, many find Christmas shopping stressful, thus helping to keep Christmas Eve the busiest shopping day of the year.[8] On the other hand, to the degree that individuals consider certain chores uniquely unpleasant, the exact bundle of procrastinating tasks will differ from person to person. Depending on the nature of their dillydallying owners, some households contain kitchen counters cluttered with dishes, while others have medicine cabinets stuffed with long-defunct prescriptions. Some have fridges needing to be filled with food, while others have dining room tables needing to be filled with friends.

---

* As determined by four separate surveys of the Procrastination Assessment Scale—Students, which assesses twenty-six possible reasons for procrastinating.

Given the connection between what is pleasurable and what is promptly pursued, it makes sense, then, that chronic procrastinators tend to detest life's allotment of responsibilities. Their jobs, their chores, their duties are all irksome, and they avoid tackling these tasks as long as possible. If you agreed with statements like "Work bores me" or "I lack enthusiasm to follow through with my responsibilities," the absence of pleasurable value is likely a source of your procrastination. Laundry makes you listless, cooking makes you crabby, and washing dishes and paying the bills are hardships rather than innocuous incidentals. You have tremendous difficulty keeping your attention on the mundane. For you, boredom signals irrelevance and your mind slides off to something else.[9] This very characteristic has provided me with quite a challenge in writing this book. I am painfully aware of your fickle nature and your unforgiving attention span—meaning that I'd better keep a lively pace at all times. In other words, ever onward.

TIME-SENSITIVE TOM

While Eddie's Expectancy and Valerie's Value are contributing factors to procrastination, Tom's reason is at its core. Tom had to book a hotel room but couldn't find the motivation until just before the deadline, letting himself get distracted every time he made an intention to act. When he finally did do something, he knew he should have acted earlier and he suffered for his tardiness. In all likelihood, if you procrastinate, you feel some kinship with Tom and have admitted that you too "get into jams" because you are "entranced by some temporarily delightful activity" or that you "choose smaller but more immediate pleasures over those larger but more delayed." The biggest factor in determining what you pursue is not the associated rewards or the certainty of receiving them, but their timing. You value rewards that can be realized quickly far more highly than rewards that require you to wait; simply, you are impulsive.

As I mentioned in the last chapter, scientific evidence of the connection between impulsiveness and procrastination is unequivocal. Scores of studies based on many thousands of people have established that impulsiveness and the related personality traits of low conscientiousness, low self-control, and high distractibility are at the core of procrastination. I myself have collected personality profiles from more than twenty thousand people to take a closer look. And I found confirmation that of these traits, impulsiveness shares the strongest bond with procrastination. This isn't surprising if you look at specific aspects of impulsiveness: intense cravings, a lack of caution and reserve, and an inability to see tasks through.[10] Though all have their role in why we put things off, the last of these is almost equivalent to procrastination in itself: not seeing tasks through means agreeing to statements like "I'm not good about pacing myself so as to get things done on time." People who act without thinking, who are unable to keep their feelings under control, who act on impulse, are also people who procrastinate.

The influence of time itself also contributes to the connection between impulsiveness and procrastination. We tend to see tomorrow's goals and concerns abstractly—that is, in broad and indistinct terms—but to see today's immediate goals and concerns concretely—that is, with lots of detail on the particulars of who, what, where, and when. Actions or goals framed in abstract terms, like "engaging in self-development," are less likely to be immediately pursued than goals framed in concrete terms, like "reading this book."[11] Similarly, the broad goal of "exercising" is less motivating than "running for an hour," and "getting a promotion" is harder to act on than the more immediate goal of "writing this report." Since we consistently frame long-term goals abstractly, the result is that we are more likely to postpone them, at least until they become short-term goals and we start thinking about them concretely. Psychologists Nira Liberman and Yaacov Trope have recently specialized in the scientific study of this phenomenon, but the basics aren't that new. David Hume wrote about the same thing over 250 years ago in his book *A Treatise of Human Nature.*[12]

Right now, if you like, you can experience the influence of time on whether you view events concretely or abstractly. Let's plan a shopping trip for the distant future, say next year. Take a moment and imagine yourself twelve months from now. What would you buy? Do you have a clear picture or is the vision cloudy and smudged? Now imagine the money currently warming your pocket. If you had to spend it today, this very moment, where exactly would that money go? Likely, what you plan to buy a year from now seems generic, as vanilla as "nice shoes" or "good sports equipment." Such goals are ghosts, ethereal and with no handles to grab onto. Today's spending plans, however, are likely concrete and meaty, something you can sink your teeth into. Instead of "shoes," they are Manolo Blahnik's "Sizzle," the python sandal that will make you the envy of every shoe fetishist. Rather than just "sporting goods," you are obsessing over a TaylorMade Quad r7 425 TP Driver, the one with the oversized titanium sweet spot, used by the pros on the PGA tour. As you contrast these concrete and abstract options, the differences in their ability to excite you should be palpable. This is procrastination's dark heart. It is largely because we view the present in concrete terms and the future abstractly that we procrastinate.

PUTTING THE PIECES TOGETHER

Eddie's, Valerie's, and Tom's situations—that is Expectancy, Value, and Time—are the basic components of procrastination.

Decrease the certainty or the size of a task's reward—its expectancy or its value—and you are unlikely to pursue its completion with any vigor. Increase the delay for the task's reward and our susceptibility to delay—impulsiveness—and motivation also dips. Understanding procrastination at this component level isn't bad, but we can do better.*

The first step is to figure out how Expectancy and Value fit together. To this end, we can tap into an entire family of formulations called Expectancy Theories, the most famous being *Expected Utility Theory.* You might not have heard of it by that name but you are more familiar with it than you know. Expected Utility Theory forms the basis of mainstream economics; every successful gambler adheres to its rule. It proposes that people make their decisions by multiplying expectancy and value together. That is:

EXPECTANCY × VALUE

Here is how it works. Imagine there are two piles of money in front of you. The one on the right I will definitely give you—very nice of me—but the one on the left I probably won't. If you could ask for only one pile, which would it be? My bet is that you would take the sure thing, revealing how *expectancy* affects your decisions. Expectancy, as you might expect, refers to probability or chance. We prefer *more* likely to *less* likely

* In fact, I wrote an article called "Integrating Theories of Motivation," dedicated to doing better. Regularly assigned reading for university students around the world, the paper acknowledges that there are a hundred years of motivational science to draw upon conducted by an army of researchers. Let's not let this go to waste.

rewards. However, what if I told you the sure thing on the right was a much smaller pile of money than the riskier one on the left? This is actually a pretty common situation, like choosing whether to put your money into riskless but low rate government bonds or to speculate on the stock market. To make sense of your options, now you have to incorporate *value* into your decision making in order to judge how much bigger the pile needs to be to inspire you to take more risk. As I vary the size of a pile and the probability of you receiving it, your preferences will flip from right to left and vice versa. The formula "Expectancy × Value" does a fair job of predicting which pile you would end up choosing. Multiplying the two together, you go with whatever pile has the highest outcome. Economists try to understand all of human behavior using just this equation. From their viewpoint, every choice you make—from pouring milk on your cereal to wiping your child's nose—is based on how much pleasure you will receive and the degree of certainty that you will receive it. Unfortunately, they are wrong.

You can't rely on "Expectancy × Value" alone to describe human nature. For starters, the theory is considered an expression of rational decision making, meaning that it doesn't leave room for any form of irrational behavior. No matter what you do, from eating an ice cream cone to getting hooked on heroin, it is all reasonable from an economist's perspective. Consequently, their theory also excludes the possibility of procrastination—irrational delays—and since I am currently writing a book on the topic and you are currently reading one, let's consider this a weakness.[13] The economic model of human nature isn't so much incorrect as just incomplete. Consistently, we do respond to incentives (i.e., value) to the extent we believe (i.e., expect) that they are obtainable, but that isn't the entire picture. There is a third factor—time.

Economists need to update how their model of human nature deals with time, and I'm not the only one saying so. Back in 1991, in a lecture aptly titled "Procrastination and Obedience," the Nobel Prize–winning economist George Akerlof spoke to the American Economic Association about how his field would be better off if it considered how we irrationally find present costs more salient than future costs. In the following year, George Loewenstein, an economics professor from Yale, co-authored *Choice Over Time,* which reviews how economics can best include time. Since then, behavioral economics, a sub-speciality of economics that integrates time, has opened up, with researchers such as Ted O'Donoghue and Matthew Rabin studying procrastination specifically. These behavioral economists are simply using observations of the world to refine their model, which is like using feedback from your eyes to keep the car on the road. Sounds very, you know . . . rational.

The theory of time that these behavioral economists are most attracted to is from the psychological field of behaviorism. Behaviorists developed a little equation called the *Matching Law,* which proved pretty good at predicting the average behavior of mice and men. Here it is in one of its simplest forms:

$$\frac{\textbf{EXPECTANCY} \times \textbf{VALUE}}{\textbf{DELAY}}$$

Since the product of Expectancy × Value is divided by Delay, the greater the delay, the less your motivation.

How important is the inclusion of time? Let me invent my own game show called *Now Deal or Then Deal.* You are a contestant and have won $1,000. It is put into your hands in ten crisp $100 bills, a short stack you stuff into your pocket. However, I also have a certified check—guaranteed money—postdated to one year from now. Here is the dilemma. What is the minimum amount that I have to put on that check to get you to dig into your pocket, hand back all those hundreds, trade me for the check, and wait 365 days to cash it? I have run this little thought experiment with hundreds of people in my classes, and most say that they would wait a year for between $2,000 and $3,000, especially if I ask for an immediate, gut decision. Unless you have been taught a reasonable rate of return and given time to mentally calculate it, thereby preventing yourself from reacting emotionally, it is likely that these responses are not so far off from your own. The more money you require to make the swap, the more sensitive you are to delay; that is, the more impulsive you are. Unfortunately, this sensitivity to delay is still missing from the equation.

Impulsiveness provides the last missing piece of the puzzle, updating the basic Matching Law. Impulsiveness provides a more sophisticated understanding of time by letting the effects of delay grow and shrink. The more impulsive you are, the more sensitive you will be to delay and the more you will discount the future—and, in the game of *Now Deal or Then Deal,* the more cash you'll require to endure waiting. Without impulsiveness, there wouldn't be such a thing as chronic procrastination. Popping this into our equation, we have:

$$\frac{\textbf{EXPECTANCY} \times \textbf{VALUE}}{\textbf{IMPULSIVENESS} \times \textbf{DELAY}}$$

And there it is: the Procrastination Equation—inspired by the common elements that determine when we procrastinate, and crafted together from the most deeply researched elements of social sciences' strongest motivational theories.* The Procrastination Equation accounts for every major finding for procrastination. As the deadline for any task gets pushed further into the future, Delay increases and our motivation to tackle the task decreases. Impulsiveness multiplies the effects of Delay, and so impulsive people feel the effects of time far less acutely, at least at first. Consequences have to be on their doorstep before they start paying attention to them—unless they are particularly large. And what makes consequences large? Expectancy and Value. The bigger the payoff and the greater the likelihood of receiving it, the sooner it will capture your attention. The Procrastination Equation also explains one of the most pernicious aspects of procrastination: the intention–action gap.

Studies show that procrastinators usually make the same plans to get to work as their more diligent counterparts. Where they differ is in acting upon their plans. Unfortunately, what was a heartfelt intention to work next week or next weekend seems a lot less important when the moment of truth actually comes around. Instead of buckling down to work, the procrastinator's intentions buckle. Unsurprisingly, one of the most common laments of procrastinators is, "No matter how much I try, I still put things off!" This complaint illustrates an intention–action gap: you truly don't want to slack off tomorrow but you constantly find yourself slacking off when tomorrow comes.

* The strict equation also includes the addition of a small constant at the bottom, typically the number "1," as in "Impulsiveness × Delay + 1." This constant's principal purpose is to prevent the equation from skyrocketing to infinity if impulsiveness or delay ever reaches zero.

This is exactly what the Procrastination Equation predicts and here's why.

Let's create an intention. Two weeks from now, you will have a choice between staying up late and honing a budget proposal for work, due the next day, or meeting your friends for drinks at the bar. At the moment, you value polishing your proposal far more than seeing your friends, as the former could lead to a sizable pay raise while the latter will only be a fun get-together. You wisely intend to work on the proposal that night, but will you stand fast? Flash forward two weeks to the very night the choice must be enacted, and life suddenly switches from the abstract to the concrete. It isn't just friends, it is Eddie, Valerie, and Tom. These are your best friends; they are texting you to come down to the bar; Eddie is so funny; Tom owes you a drink and you owe Valerie a drink; and maybe you can bounce some ideas off them. Besides, you deserve a break because you've worked so hard. So you give in, and once you are there, you forget about going back to work. Instead, you pledge to get up early tomorrow morning because "your mind will be fresher then." The culprit for your intention–action gap is time. When you headed down to the bar, it probably took you just 15 minutes to get there, a minuscule delay compared to the deadline for tomorrow's task, which is orders of magnitude off into the future—specifically 96 times greater (i.e., 24 hours divided by 15 minutes). As per the Procrastination Equation, that difference causes an almost hundredfold increase in the relative effects of delay. Indeed, there's no time like the present, and it's no wonder your intentions fell through.

THE PROCRASTINATION EQUATION IN ACTION

To see all the pieces of the Procrastination Equation in action at once, it would be tempting to try swapping in your own scores on impulsiveness, expectancy, and value and checking out the results. Unfortunately, it isn't that easy. To accurately apply the equation to a specific individual we would need a controlled laboratory experiment. In the lab we can put everything into an exact and measurable metric by artificially simplifying your choices, having you push a bar, or run a maze to receive a food pellet, for example.

To demonstrate how the Procrastination Equation operates in a realistic setting, a better way is to apply it to the prototypical procrastinator. And nobody—nobody at all—procrastinates like college students, who spend, on average, a third of their days putting work off. Procrastination is by far students' top problem, with over 70 percent reporting that it causes frequent disruption and fewer than 4 percent indicating that it is rarely a problem.[14] Part of the reason that colleges are filled with procrastinators is that their inhabitants are young and therefore more impulsive. However, the campus environment must shoulder most of the blame. Colleges have created a perfect storm of delay by merging two separate systems that contribute to procrastination, each devastating in its own right.

The first system is the essay. The more unpleasant you make a task—the lower its value—the less likely people will be to pursue it. Unfortunately, writing causes dread, even revulsion for almost everyone. But welcome to the club. Writing is hard. George Orwell, author of the classics *Nineteen Eighty-Four* and *Animal Farm,* had this to say: "Writing a book is a horrible, exhausting struggle, like a long bout of some painful illness. One would never undertake such a thing if one was not driven by some demon that one can neither resist nor understand." Gene Fowler, who wrote about twenty books or screenplays, was equally despairing: "Writing is easy, all you do is sit staring at a blank sheet of paper until the drops of blood form on your forehead."

To write this very book, I have been leaning on William Zinsser's *On Writing Well: The Classic Guide to Writing Nonfiction.* Sure enough, on page 87, Zinsser confesses, "I don't like to write."

Added to the cruelty of assigned writing is the capriciousness of grading—low expectancy. Any essay that is re-marked by another professor may shift remarkably in grade—a B+ could become an A+ if you are lucky, or a C+ if you are not.[15] This is not because the marker is sloppy; it is because measuring performance is inherently hard. Just look at the variation in judges' scores at Olympic events or among reviewers critiquing films. From the students' perspective, such discrepancies mean there is no guarantee that their hard work will be recognized. Quite possibly, it won't be.

The final aspect of the essay system that contributes to student procrastination is the distant due date—high delay. There are often no intermediary steps—you just hand the paper in when you are finished. At first the due date seems months and months away, but that is just the start of a slippery slope. You blink and it has become weeks and weeks, then days and days, and then hours and hours, until suddenly you are considering Plan B. Approximately 70 percent of all reasons given for missing a deadline or bombing an exam are excuses, because the real reason—procrastination—is unacceptable.* As students themselves report, their top strategy is to pore over the instructions with a lawyer's eye for any detail that could possibly be misinterpreted, later claiming, "I didn't understand the instructions."[16]

There it is; university essays hit each key variable of the equation. Essays are grueling (low value), their results are very

* The most infamous excuse is to claim the death of a grandparent. Mortality of grandparents increases several hundredfold during final exams, a statistic that if taken seriously suggests that seeing the grandkids being tested is extremely stressful for the elderly.

uncertain (low expectancy), and they have a single distant deadline (high delay). And if essays are hard in and of themselves, there are few harder places to do them than a college dorm. This leads us to the second system in that perfect storm: the place where this essay is supposed to be written.

College dorms are infernos of procrastination because the enticements—the alternatives to studying—are white hot. Superior in every aspect to essay writing, these pleasures are reliable, immediate, and intense. Consider campus clubs alone. At the university where I earned my PhD, there are about a thousand of them, catering to every recreational, political, athletic, or spiritual need, ranging from Knitting for Peace to the Infectious Disease Interest Group. These clubs will give you a new set of friends, with whom you will want to socialize—likely in one of the dozens of coffee shops and pubs a short walk from any place on campus. They'll also entice you to go to one of a dozen events occurring every week, from poetry readings to tailgate parties. With all the camaraderie, alcohol, sex, and—headiest of all temptations—the freedom to enjoy them all, university can lure us into the unregulated state of bliss where the liberties of adulthood are combined with only a minority of the responsibilities. From the moment students step into the classroom, inevitable conflicts are set in motion. Even Tenzin Gyatso, better known as the 14th Dalai Lama, reported of his student days, "Only in the face of a difficult challenge or an urgent deadline would I study and work without laziness."

We can graph this dilemma using Eddie, Valerie, and Tom when they were back in their university days. They hang together, as they have a lot in common and all of them like to socialize rather than work. Still, there are differences among them. Valerie knows she isn't especially bright but she has two cardinal virtues—she is levelheaded and responsible. Though she isn't competitive, she sees the future pretty clearly, and can imagine one day graduating from college and getting her dream job. Tom is more ambitious and more confident of his abilities than either of his classmates but he is also the most impulsive. His cockiness and spontaneity arouse mixed feelings of envy and hate in many people who know him. Eddie, on the other hand, lacks desire as well as self-confidence. He was pressured by his family to go to college and he is unsure whether he can survive much less thrive academically. In fact, he doesn't really care. At least he is comfortable being a slacker.

One mid-September morning, Eddie, Valerie, and Tom walk into my Introduction to Motivation class, where they find that a final essay is due three months later, on December 15. The graph on the facing page charts their likely levels of motivation and when each of them will start working. Their common motivation to socialize, represented by the dotted line, starts off strong in the semester and tapers off toward the end, partly in response to a lack of opportunity and ever-increasing guilt. Valerie, being the least impulsive, is the first to start working, on November 29 (the smooth, unadorned graph line). It takes another week before Eddie or Tom bears down—a significant gap.

In terms of the Procrastination Equation, although Tom is more confident (high expectancy) and competitive (high value) than Eddie, his impulsiveness means that most of his motivation is reserved until the end (the graph line with squares). Valerie's motivation flows more steadily, like water from a tap, while Tom's gushes like a fire hose when eventually turned on. Even though Tom starts working the same day as Eddie the slacker (the graph line with triangles), Tom's motivation in the final moments should enable him to outstrip the others' best efforts.

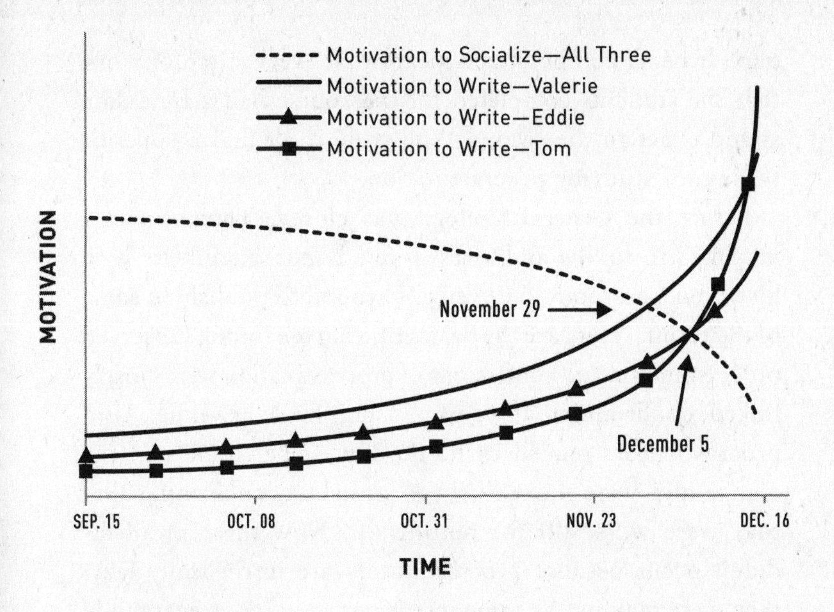

MY OWN RESEARCH

Although Eddie, Valerie, and Tom are fictional, they are composite characters based on the thousands of students I have taught. As I stressed, there is no better venue for finding procrastinators than universities. Harnessing all this wasted motivation for science is the trick. It was great luck that as a graduate student I worked with Dr. Thomas Brothen. Thomas taught an introductory psychology course at the University of Minnesota's General College, an institution designed specifically to increase the diversity of the university. Significantly, the class was administered through a Computerized Personalized System of Instruction, a nifty arrangement that allows students to progress through a course at their own pace but is well known for creating high levels of procrastination. In fact, procrastination is such a problem that students are repeatedly warned throughout the course about the dangers of delay. And here is the beautiful part. It being computerized meant that every stitch of work that the students completed for the course had a time-date stamp exact to the second. You truly can't find a superior setting for studying procrastination.

Before the General College was closed, Thomas and I managed to follow and assess a few hundred students with his wired classroom. We even got around to publishing some of the results. Here are the basics of what we found. Observed procrastination and confessions of procrastination were closely linked, confirming that we were using the right venue. Also, procrastinators tended to be the lowest performers in the course and were more likely to drop out, confirming that they were worse off for putting off. Now these problems didn't occur because procrastinators are intrinsically lazy; they were making the same intentions to work as everybody else. They just had trouble following through with their intentions at the beginning of the course. Toward the end, a different story emerged. Procrastinators actually started logging more hours than they had intended, with one student completing 75 percent of the course in the final week alone. They also weren't procrastinating because of anxiety. The real reasons for inaction were the following: impulsiveness, hating the work, proximity to temptation, and failing to plan. And most significantly, each of these findings directly follows from the Procrastination Equation.

The ability of the Procrastination Equation to formulate these and other results forms the backbone of this book. I have already talked in depth about the connection between the intention–action gap and impulsiveness. Similarly, putting off work because it is unpleasant simply illustrates the effect of value on procrastination. Proximity to temptation highlights the effect of time. Students who said that if they chose not to study "they could immediately be doing more enjoyable activities" or that in their study location there were "a lot of opportunities to socialize, to play, or to watch TV" procrastinated more, a lot more. Remember that Eddie, Valerie, and Tom needed their motivation to write to exceed their motivation to socialize before they could get down to work. But the more readily available temptations become, the stronger they become and the longer they will dominate choices, necessarily creating procrastination. The findings from our study, such as procrastinators' failure to properly plan or to create efficient study schedules, also pointed to ways of combating procrastination. Proper planning allows you to transform distant deadlines into daily ones, letting your impulsiveness work for instead of against you. We will talk more about how to plan properly and the rest of these issues as you go through the book. But one last thing about this study.

There is an epiphany I want to share that occurred to me when I graphed the work pace of the class. Would their work pace replicate the curve that the Procrastination Equation predicted, starting off slow and then spiking toward the end like a shark's fin? Would it follow the pattern that Eddie, Valerie, and Tom's experience suggested? I couldn't expect an exact match, as the equation couldn't take into account weekends or the midterm-break lull, but I was hoping for something close. My findings are what you see on the next page. The dotted line is a hypothetical steady work pace, the dark line is what we observed, and the gray line is what the Procrastination Equation predicts. Notice which lines match together almost perfectly.[17]

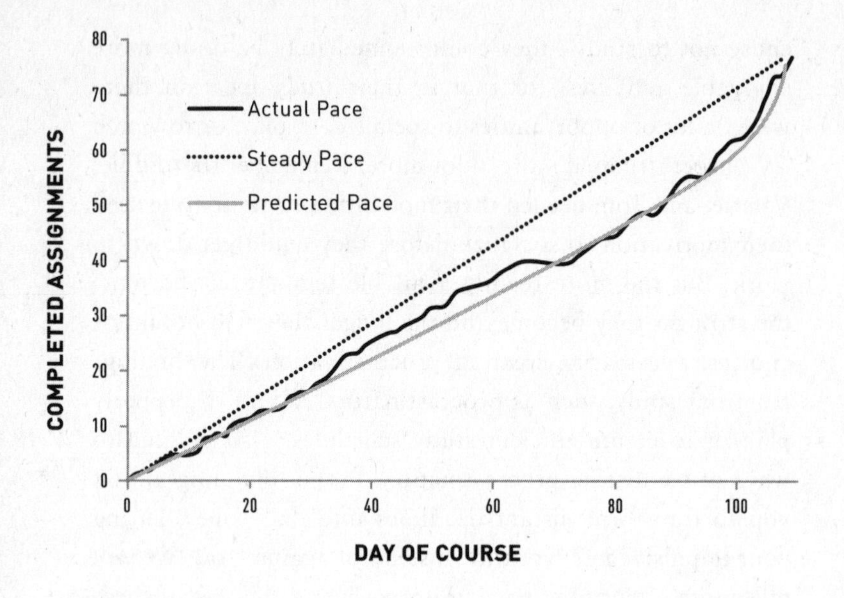

LOOKING FORWARD

To some, a mathematical model of procrastination is threatening; it reduces humankind to a robotic formula. I am sympathetic. We are all more complicated and nuanced than any equation could capture, and the subtle details of each person's procrastination are personal. Exactly when your self-confidence peaks, what you find deathly dull, and where your vices lie all combine to determine your individual procrastination profile. The Procrastination Equation isn't seeking to form a comprehensive depiction of who you are but to create a succinct snapshot that can explain a lot with a little.

The Procrastination Equation attempts economically to describe the underlying neurobiology that creates procrastination. I will tell you right now: the biology and the math won't match exactly. A road map of a city, for example, no matter how recent or detailed, can't represent every corner and crevasse of reality; it skips over details like architectural styles or fire hydrant placement. Judiciously focusing on streets and highways allows the map to emphasize navigation. If this big picture doesn't satisfy you and you want all the details, don't fret. The next chapter will give you what you are looking for.

CHAPTER THREE

Wired for Procrastination

PUTTING OFF IS HUMAN NATURE

Think of all the years passed by in which you said to yourself "I'll do it tomorrow," and how the gods have again and again granted you periods of grace of which you have not availed yourself. It is time to realize that you are a member of the Universe, that you are born of Nature itself, and to know that a limit has been set to your time.

MARCUS AURELIUS

Every day, we experience our souls as being split.[1] Who hasn't struggled between a reasonable intention and an opposing pleasurable impulse? As the dessert cart pulls into view, commitment starts to crumble in the heat of the internal battle of "I want to eat that cake, but I don't *want* to want to eat cake." Have you ever skipped exercising, knowing that you would later regret it? Have you ever scratched an itch, knowing that you just made it worse? You are not alone; it's a permanent part of the human condition. Thousands of years ago, Plato described this internal clash as a chariot being pulled by two horses, one of reason, well-bred and behaved, and the other of brute passion, ill-bred and reckless. At times, the horses pull together and at other times they pull apart. Thousands of years later, Sigmund Freud continued Plato's equestrian analogy by comparing us to a horse and rider. The horse is desire and drive personified; the rider represents reason and common sense. This division has been rediscovered by dozens of other investigators, each with their own angle, emphasis, and terminology for the same divided self: emotions versus reason, automatic versus controlled, doer versus planner, experiential versus rational, hot versus cold, impulsive versus reflective, intuitive versus reasoned, or visceral versus cognitive.[2] Understanding how the architecture of the brain enables this division is the secret to understanding the biological basis of procrastination.

The brain has been considered the last frontier of human science because its workings have been so difficult to investigate. Emerson Pugh, a Carnegie Mellon University physics professor, concluded that, "If the human mind was simple enough to understand, we'd be too simple to understand it." He is right. And the Procrastination Equation is only a model of how you might behave. Though I like to think of it as a supermodel, it is still merely an approximation of how motivation works. Our brains aren't actually doing these calculations any more than a falling stone is calculating its mass times its acceleration to determine with what force it will hit the ground.[3] Rather, the equation summarizes a more complex underlying process, the interplay between the limbic system and the prefrontal cortex. This is where we must turn for a more fundamental understanding of procrastination.

Recent advances in brain science have allowed us to pull the curtain aside and see our own minds in operation. The basic methodology isn't that hard to describe. You place participants in your choice of brain scanner, likely a functional magnetic resonance imager (fMRI), which detects subtle changes in magnetic signals associated with blood flow and neural processing (i.e., thinking). Once the participant is strapped in, you then ask questions carefully designed to engage aspects of decision making and observe which parts of the brain light up. For example, if we had J. Wellington Wimpy as a subject, we could ask him, "If I gave you a hamburger today, how much would you pay me on Tuesday?" Sure enough, what then comes up on the electronic monitors are not one but two internal messages, which science has blandly come to call System 1 and System 2.[4]

Asking a thirsty person a question such as what drink she would like *now* primarily activates System 1, the limbic system. This is the beast of the brain ("the horse"), the origin of pleasure and of fear, of reward and of arousal. Questions about future benefits, however, activate System 2, the prefrontal cortex ("the rider"). Though studies are still refining the exact subsection of the prefrontal cortex that is involved, the consensus is that this is willpower's throne. The prefrontal cortex is often described as the *executive* function, appropriately evocative of CEOs making strategic company plans. Without it, long-term pursuits or considerations become almost impossible, as it is—literally—what keeps our goals in mind.[5] This prefrontal cortex is the place from which planning arises. The more active it becomes, the more patient we can be. It allows us to imagine different outcomes and, with help from our speedy and definitive limbic system, helps us to choose what to do. This interplay of instinct and reason has enabled the human race to create the world in which we live. But it also has created procrastination.[6]

You see, this decision-making arrangement is not the most elegant. It's often described as a haphazard *kluge,* the clumsy outcome of an evolutionary process.[7] Because the limbic system evolved first, it is very similar across species. It makes decisions effortlessly, spurring action through instinct. Its purview is the here and now, the immediate and the concrete. Our more recently evolved prefrontal cortex is more flexible in its decision making, but also slower and more effortful. It is best at big-picture thinking, abstract concepts, and distant goals. When the limbic system is aroused by immediate sensations of sight, smell, sound, or touch, an increase in impulsive behavior results, and the "now" dominates. Future goals espoused by the prefrontal cortex are cast aside and we find ourselves seduced into diversions—despite knowing what we should be doing, we simply don't want to do it. Also, because the limbic system runs automatically at an incredibly fast rate and is thus less accessible to consciousness, desires can often come over us inexplicably and unexpectedly.[8] People feel helpless to stop intense cravings and they display little insight into their ensuing actions other than, "I felt like it."

In essence, procrastination occurs when the limbic system vetoes the long-term plans of the prefrontal cortex in favor of the more immediately realizable; and the limbic system, aside from being the quicker of the two and in charge of our first impulse, is often the stronger. When near events get this evaluative boost from our limbic system, their vividness increases and our attention shifts to their immediate and highly valued consumable aspects (what we can see, smell, hear, touch, and taste). Deadlines are often put off until they are close or concrete enough to get a hint of that limbic system zing, whereupon both parts of our brain are finally shouting in agreement, "Get to work! Time is running out!"

OF BABES AND BEASTS

Procrastination increases whenever our more recently acquired prefrontal cortex is compromised.[9] The less potent the prefrontal cortex, the less patient we become.[10] Those with brain damage can provide particularly vivid examples of this, Phineas Gage being the most famous.[11] Gage was a shrewd, responsible, hardworking, methodical railway foreman who, in a workplace accident in 1848, had over three feet of iron rod blown through the top of his skull and the front of his brain. He recovered, incredibly, but he became a man of the moment: impatient, vacillating, profane, inconsiderate, uninhibited, and uncontrollable. The iron rod had severed the connection between Gage's limbic system and his prefrontal cortex. The planning part of the brain needs the fast and accurate input from the limbic system to understand the world, and that's what Gage lost. A more modern example is Mary J. who was completely transformed within a year by a brain tumor that debilitated her prefrontal cortex.[12] Before the tumor, she was a quiet teetotaling Baptist, on the dean's list at an Ivy League university, and engaged to be married. Until the tumor was surgically removed, she was angry and extremely promiscuous, failing school, drinking hard, and using drugs. Her executive function was disabled and she became all impulse, ruled by whatever temptation was put before her.

There is a way people can experience Phineas' or Mary's predicament, and happily it doesn't involve a nail gun. We can temporarily lesion the prefrontal cortex with transcranial magnetic stimulation, which uses electromagnetic induction to briefly knock out focused sections of the brain.[13] Alternatively, taking alcohol, amphetamines, or cocaine either hypercharges the limbic system or hinders the prefrontal cortex's ability to perform, creating actions that "seemed like a good idea at the time" but later prompt regrets.[14] Or, the prefrontal cortex can simply become exhausted through sleeplessness, stress, or resisting other temptations; by fighting one enticement, we often become more susceptible to another.[15] Finally, if you are a teenager, you might not need to go to any of these extremes, since your prefrontal lobes are still receiving their final touches.[16] Combining the effects of youth, stress, and alcohol together, the most impulsive and uninhibited place on this planet is a group of teenagers celebrating the finish of a willpower-depleting stretch of studying with a weeklong drink-fest. Indeed, Phineas Gage would fit right in during Spring Break in Cancún, with wet T-shirt contests, drinking games, and random hook-ups. If you don't diminish the prefrontal cortex, you can't have *Girls Gone Wild.*

If you can't make it to Spring Break to see the limbic system dominate action, there are good alternatives closer to home. In fact, they are likely in your home. Do you have a pet or a child? Both are heavy on the limbic system, making owning a pet the neurobiological equivalent of raising a child.[17] We are their external prefrontal cortices. We have to be the ones providing patience and doing our best to coax it out of those who don't have much of it or who are still developing it.

THE NOW OF BABES

There is a rhyming biological heuristic that goes "ontogeny recapitulates phylogeny." It means that the way we develop within our lives roughly reenacts the course that human evolution has taken over millions of years. When in the womb, more or less, we morph from fish to reptile before eventually emerging as a mammal. But the process isn't done yet. The last aspect of us to evolve is the prefrontal cortex, which continues to develop after birth.[18] For those who have children, and as I write this, I have two still in diapers, we don't need a biology degree to know that infants aren't born with the ability to plan ahead and put their immediate needs on hold for the benefit of some future goal. Just try asking for patience from a hungry baby or a little one with a full diaper and my point will be made. They are merciless in their need.

As children develop, their prefrontal lobes grow too and eventually they achieve the ability to put things off just a bit. You can't ask a baby to put off a feeding, but eventually you may ask a toddler to say "Please" before getting a treat. It takes the development of the prefrontal cortex for this modicum of control to appear—which happens all too slowly for my taste. Year-old children have almost no executive control, instantly batting down any pile of blocks or grabbing your eyeglasses, but just one year later, brief moments of patience become possible, say twenty seconds. By the age of three, children are routinely waiting a full minute and by four they are piling their blocks high, putting off the blast until they can enjoy the big burst of noise when their soaring towers tumble.

At the age of four, children can play "Simon Says." This is a significant advance, because the game is all about self-control, about inhibiting the immediate impulse from the limbic system so that the prefrontal cortex can mull over whether Simon has actually said "Simon Says" before they respond. How well this acquired ability transfers into kindergarten is another matter, because kindergarten requires sitting when you want to run, listening when you want to shout, and taking turns when you want it all to yourself. Fortunately, between the ages of four and seven there is a burst of development in children's executive function. They are progressively better able to make plans for tomorrow, to pay prolonged attention to more than the television set, and to shut out distracting events other than parents calling them to come in for dinner.

The normal maturation of the prefrontal cortex is assisted by endless hours of patient teaching by parents trying to get their little ones to put off their needs for just a moment without tears or the stomping of feet. Unwearyingly insisting that gifts can be unwrapped only at Christmas and then only your own, that dessert comes after dinner, or that toys must be shared with others coaxes a little more from the prefrontal cortex and a little less from the limbic system. Unfortunately for parents, their role as their children's external prefrontal cortices is a long one. It can last until about the age of nineteen or twenty, when the biological basis of self-control is finally fully in place. Until then, parents can only herd their teenagers away from all the vices that impulsiveness ensures youth will find especially tempting: risky sex, excessive alcohol, petty crime, reckless driving, and, of course, procrastination.[19] The younger you are, the more you seek instant gratification, from socializing late into the night and then facing tomorrow's exam half asleep to dillydallying so long you have to pack your bags in a flurry and almost miss your plane. Though the young act as if they will live forever, they really are living just for today.

The novelist Elizabeth Stone has written that having a child "is to decide forever to have your heart go walking around outside your body," but our role as walking prefrontal lobes comes to an end at this point. As adults, our children no longer need us for guidance and any mental inequalities between us go into a long lull, perhaps broken briefly by the arrival of grandchildren if they are forthcoming. We can expect apologies from our kids as they try to raise a few of their own and learn firsthand the vigilance required to be a parent. And then, long later and hopefully not at all, our roles may change entirely. As we grow older so do our brains, losing the snap they had in earlier years, especially the prefrontal cortex—following the last in, first out rule.[20] Though some avoid this fate, remaining razor sharp into their final years, others get it worse, assisted by the senility of frontotemporal dementia that affected my grandmother Eileen.[21] I am well aware that I too might encounter a second childhood and once again be as vulnerable as my two young sons are now. Indeed, we'd better raise our kids well, as their love might be the only thing that stands between us and a world that views us as prey made easy through old age and a compromised mind.

BIRD BRAINS

Animals might be our fellow procrastinators. After all, we share many other "human" personality traits with dozens of other species, from rhesus monkeys to octopi. Wild great tits, for example, exhibit varying degrees of aggressiveness and risk taking, traits that enable greater environmental exploration.* The bolder birds expose themselves to more danger but also reap the gains of better nesting places, food sources, and choice of mates.[22] For another example, just ask any dog or cat owner if their pets have a unique personality; the owners will rightly insist that their furry friends differ in terms of affection, anxiousness, aggressiveness, and curiosity.[23] Significantly, this list of shared traits includes impulsiveness, the cornerstone of procrastination.[24] But this doesn't necessarily translate into procrastination itself.

Whether they are meowing, barking, or chirping, animals are clearly limbic-heavy in their decision making. But that's only half the story. You need some prefrontal cortex or its

* The Latin name for Great Tits is *Parus major*; despite the suggestive name, all but begging to be made fun of, they are the best studied bird in the world.

equivalent to procrastinate, for without it you can't make plans that you later irrationally put off. Do animals have this mental capacity? Apparently some do, showing the ability to anticipate and plan for the future, especially regarding food.[25] Scrub jays can anticipate being denied breakfast tomorrow and will cache food to snack on later. Rats seem to have some sense of time, being able to recall where and when feeding events occur.[26] Chimpanzees can wait up to eight minutes to exchange a small cookie for a large one, showing slightly more patience than a young human child.[27] Male chimps will invest in future mating opportunities by sharing meat with a female, with the hope of being favored when she comes into heat.[28] Also, consider Santino, a particularly farsighted chimpanzee from the Swedish Furuvik Zoo. He will spend his morning collecting stones to hurl at annoying zoo visitors in the afternoon.[29] In combination with impulsiveness, all the pieces for procrastination are there: animals can make plans for the future and, what's more, they can impulsively put them off, despite expecting to be worse off for it.

James Mazur, a Harvard-trained psychologist, has directly demonstrated procrastination in animals. He trained pigeons to two different work schedules and then gave them a choice of which to pursue. Both schedules delivered a tasty treat at the same time, but the first started with a little work followed by a long delay, while the second started with the long delay and ended with *a lot* more work, up to four times as much. Essentially, the birds had to choose between doing a little hard work now (followed by rest and recreation) and taking it easy immediately (followed by a lot of hard work). The pigeons proved to be procrastinators, putting off their work despite having to do more of it to obtain their reward in the end.[30] Like a twisted version of a Cole Porter song, birds delay doing it and even chimpanzees in the zoo delay doing it. Since most animals, including pigeons, have the capacity for procrastination, procrastination is pretty well confirmed as a fundamental part of our motivational firmament.[31] The last time we all went to the same family reunion was over 286 million years ago during the Carboniferous period, before the time of the dinosaur.

Inevitably, then, having an animal as a pet is largely an exercise in dealing with this limbic-heavy decision making. Dogs, for example, naturally act in the moment and grab food that isn't theirs, chase stray animals across busy streets, and bark or whimper incessantly by the door waiting for you to open it. It would be easier in the short run to let the dog be, but patience and long-range thinking on our part can make all the difference for a life with any four-legged friend. This is what expert dog trainers stress, like Cesar Millan, the dog whisperer, or Andrea Arden, author of *Dog-Friendly Dog Training:* the primary responsibility of an owner is "to convince your dog that waiting for something—which is typically not a natural instinct for dogs—is the best option."[32] The big trick is convincing owners to do this in the first place. Teaching impulse control uses a lot of our prefrontal cortex, a resource we often don't have a surplus of to begin with.

EVOLUTIONARY PROCRASTINATION

By all appearances, from the evidence of brain science to animal studies, the capacity to procrastinate is engrained in us. It's even in our genetic code: several studies indicate that about half of most people's lack of self-discipline has a genetic origin.[33] This makes sense, given that DNA allows adaptive genetic mutations to be passed down through subsequent generations, a process known as "descent with modification." Without a genetic component, the ability to procrastinate couldn't easily be passed on.

We evolved to be procrastinators, but why? Procrastination is an irrational delay, whereby we voluntarily put off tasks until later despite expecting to be worse off for the decision. By definition, procrastination is harmful and should have been culled long ago from our gene pool rather than filling it to the brim. Are we the butt of some cosmic joke? Maybe. But there is another possibility to consider. Some traits occur as by-products of other once-more-adaptive processes. For example, belly buttons are a by-product of being born, and though they can be pretty, they don't have any pressing purpose in themselves. Since procrastinators are above all else impulsive, the evolutionary explanation for impulsiveness is the one to focus on. Procrastination is a by-product.[34]

Essentially, impulsiveness is about living for the moment. Long-term desires and tomorrow's deadlines are ignored until they become imminent—until the future becomes the now. Though today impulsiveness isn't usually a helpful trait, evolution operates through hindsight; that is, it custom fits us to the environment we *were* in, with no anticipation or prediction. This is known as *ecological rationality,* in that what is rational depends upon the environment you operate in. It is like getting a tailored suit for your wedding day. You look magnificent in it, but try it on again twenty years later and it pinches in all the wrong places. Likewise, procrastination may be steeped within our existence because having an impulsive mindset made a lot of sense when we were hunter-gatherers. When our ancestors needed to do the basic four "F"s of survival—feeding, fighting, fleeing, and mating—it would aid their cause if they also wanted to. Let's briefly consider the last and first of these four: what we have for dinner and who we seek to spend the evening with afterward.

FAST FOOD

From our teeth, which chew it up, to our intestines, which digest it, food has played a major role in our evolution. We have evolved to love the taste of fats and sugars because, in a world where starvation and predation were constant concerns, stocking up on high-caloric foods was once an adaptive preference. When the food supply was sporadic, we would have to gorge when the going was good, focusing on energy foods rich in sugar and fat. There were no Neanderthals on self-imposed diets. Consequently, for most of human history, being "overweight" has been considered beautiful, affluent, and enviable.[35] The exigencies of eating may explain how we all became so impulsive and, consequently, procrastinators.

Let's consider two types of primates, common marmosets and cotton-top tamarins, which are almost identical except in their choice of food.[36] Marmosets are gummivores, which scratch tree bark and then sip on the sap that flows. Tamarins are insectivores; they pounce on bugs whenever they can find them. Marmosets show a lot more self-control than tamarins, as they are selected for it. Sap takes a while to flow, demanding patience, whereas the hunt for jumping and scurrying bugs requires immediate action. For animals in general, the fine tuning of impulsiveness to their food source is called *optimal foraging*.[37] We are optimized to get the most calories in the shortest time; consequently, the longer it takes to kill, eat, and digest, the less impulsive a species typically becomes. In short, we develop the self-control we need to ensure our next meal.*

---

* Gary Marcus, a New York University psychologist and author of *Kluge: The Haphazard Construction of the Human Mind,* concludes that "over hundreds of millions of years, evolution selected strongly for creatures that lived largely in the moment."

Being omnivores and at the top of the food chain, humans are superstars of self-control. We have the patience to kill and eat almost anything that lives. Birds' ability to delay gratification, in comparison, hardly registers; even a ten-second wait is remarkable. Similarly, ten minutes of patience is an eternity for a chimp. For all our superior self-control, though, in today's whirlwind, we don't have enough. We have been favored with enough patience for a world without grocery stores or refrigerators, enough for hunting animals or gathering berries. Yet, we have a relatively small window compared to what we currently need. Procrastination results from a disconnect in our genetic inheritance, as we now pursue projects and plans that require weeks, months, and years to complete, timelines for which we are motivationally mismatched. In the forest, a bird in the hand might be worth two in the bush, but in the city, the discount rate is far more slender; invest in a bird today and tomorrow you are lucky to earn a chicken wing's worth of interest.[38]

JUST SAY YES

Now on to the second example, the one you've been waiting for—sex. Evolution is steeped in sex, as those who succeed breed. Since procrastinators' impulsive nature is ingrained in their DNA, it can be passed on to their offspring and, if it lets them have more kids, the trait quickly becomes common. Just consider my family. The males on my mother's side tend to have children later in life. My great grand-dad was Owen Owen, who people in the UK might remember from his string of similarly named though now-defunct department stores.[39] Since Owen Owen was born in 1847 and I had my son Elias in 2007, each generation of my family tree is spaced forty years apart. If we were in a stork race with another family that started a new generation every twenty years (thereby reproducing twice as fast), by now there could easily be over eighty of them for every one of us. Getting an early start on baby making makes a big difference.

Sure enough, procrastinators' impulsiveness has been linked to an early start for parenthood through teen pregnancy as well as sexual promiscuity.[40] The one thing that procrastinators don't tend to put off is "getting some." No wonder. The fun part of copulation comes immediately, while the harder part of child raising . . . well, that's almost a year away. This state of sexual affairs also helps explain why men tend to be more impulsive and procrastinate more than women.[41] Reproduction strategies favor a quality versus quantity split—that is, raising a few kids well or having lots in the hope some of them work out. Since it is easier for men to invest less in their offspring, they definitely lean toward the quantity option. As Geoffrey Miller, author of *The Mating Mind,* wrote: "Men are more motivated to have short-term sexual flings with multiple partners than women are." Women tend to favor the quality strategy, taking a longer-term and more responsible perspective. She waits patiently for Mr. Right while he impulsively wants Ms. Right Away.

Sex also ensures a range of impulse-driven procrastination in the populace; some will procrastinate a little and others a lot. If it was always advantageous to get pregnant as soon as possible, the world would be like the Mike Judge movie *Idiocracy.* In that film, everyone who was smart and cautious held off having kids, and the intelligent were out-bred by the clueless and carefree. There is, however, no optimal level of impulsiveness to maximize the number of your descendants.[42] Much depends on the resources available to raise children, for as costs increase, it is better to have smaller families.[43] Other tradeoffs occur when there is an excess of men pursuing the "quantity" reproduction strategy. If too many men are focused on short-term sexual encounters, they swamp the singles bars and strain the goodwill of the available women. In this scenario, committed family men are a rarity and thus more valued. Men demonstrating loyalty would find themselves vigorously pursued, able to pick the prettiest and most compatible of spouses.

A BRIEF HISTORY OF PROCRASTINATION

This evolutionary explanation of procrastination directly demonstrates why procrastination is so widespread. No matter which country or language you are reading this book in, there is a name for irrationally putting things off, from Hawaii's *napa* to Scotland's *maffling*. Everywhere we have looked for procrastination, we've found it—easily. Today's age of procrastination was inevitable the moment we walked out of the trees into the savanna, learned to make fire, and began trading among tribes. Procrastination grew alongside civilization.

The history of procrastination likely began around nine thousand years ago, sprouting along with the invention of agriculture.[44] Planting crops in the spring to reap them in the fall was our first artificial deadline; it was a task that civilization and survival required but not one we had evolved to perform. This is why all the earliest written records of procrastination deal with farming. Four thousand years ago, ancient Egyptians chiseled at least eight hieroglyphs to indicate delay, but one in particular also indicates neglect or forgetfulness.[45] Translated as procrastination, this hieroglyph is most often associated with agricultural tasks, especially those connected with the yearly cycle of the river Nile, as it overflowed its banks and fertilized the floodplains. Similarly, the ancient Greeks struggled with procrastination, as recounted by Hesiod. Living around 700 B.C., Hesiod was one of the greatest poets of Greek literature, rivaled only by

Homer. In Hesiod's epic 800-line poem, *Work and Days,* he exhorts: "Do not put your work off till tomorrow and the day after; for a sluggish worker does not fill his barn, nor one who puts off his work: industry makes work go well, but a man who puts off work is always at hand-grips with ruin." This warning was especially important because the Greeks were in the midst of a financial crisis of such proportions that many Greek farmers put up not only their farms but also their families as collateral. Procrastination led not only to a poor credit rating, but also to seeing your sons and daughters become the exclusive property of your richer neighbors.

By 440 B.C., procrastination was spilling over from farming into fighting. Thucydides, the father of scientific history, wrote about it in the *History of the Peloponnesian War,* which chronicled the conflict between Athens and Sparta. This account, still studied in military colleges, discusses various aspects of personalities and strategies. Thucydides clearly considered procrastination to be the most wicked of character traits, useful only to delay the commencement of a war so as to lay a better groundwork for winning it. Another notable Greek reference to this trait is found in the work of the philosopher Aristotle, who wrote much of his *Nicomachean Ethics* on the weakness of the will, what the Greeks called *akrasia*. Specifically, Aristotle discusses a form of *akrasia* called *malakia,* which is not doing something that you know you should (clearly, procrastination).*

Moving a few centuries forward, we see procrastination entering politics. Marcus Tullius Cicero was a major political player around 44 B.C. His position put him in conflict with

* In Greek today, *malakia* has a somewhat fouler meaning, possibly best translated as "wanker."

Marcus Antonius, better known as Mark Antony, Cleopatra's lover. In a speech denouncing Mark Antony, Cicero declares: *In rebus gerendis tarditas et procrastinatio odiosae sunt* ("in the conduct of almost every affair slowness and procrastination are hateful"). Perhaps because of Cicero's advice, or perhaps because Cicero made thirteen other speeches denouncing him,[46] Mark Antony delayed little in killing him.

Then, for a millennium and a half, procrastination made inroads into religion and it is referenced in the texts of every major faith. For example, in the earliest written Buddhist scriptures from the Pali Canon, the monk Utthana Sutta concludes that "Procrastination is moral defilement."[47] Moving forward seven centuries, the Indian Buddhist Shantideva is still on message, writing in *The Way of the Boddhisattva,* "Death will be so quick to swoop on you; Gather merit till that moment comes!" By the sixteenth century, procrastination starts appearing in English texts without translation. The playwright Robert Greene, for example, wrote in 1584, "You shall find that delay breeds danger, and that procrastination in perils is but the mother of mishap."

Finally, when the Industrial Revolution came into its own, so did procrastination. In 1751, Samuel Johnson wrote a piece for the weekly periodical *The Rambler,* describing procrastination as "one of the general weaknesses, which, in spite of the instruction of moralists, and the remonstrances of reason, prevail to a greater or less degree in every mind."* Four years later, Dr. Johnson enshrined the word within his influential

* Naturally, Dr. Johnson procrastinated writing that very article until the last possible moment, composing it in Sir Joshua Reynolds' parlor while the errand boy waited outside to bring it to press. "Typical," as his friend Hester Piozzi remembered, given that "numberless are the instances of his writing under immediate pressure of importunity or distress."

English dictionary and, ever since, the term has remained in common use. If procrastination is indeed a core characteristic of humankind, it is acting just as you would expect: it's maintaining itself as a reoccurring theme in our history books, right from the beginning of the written word.

LOOKING FORWARD

I'd like to end this chapter on the evolution of procrastination with the story of Adam and Eve. They lived in the Garden of Eden, naked and unashamed, fitting in perfectly with nature. Then, in humankind's first act of disobedience, Adam and Eve bit an apple from the tree of knowledge, and they were cast out by God, forced to survive through agriculture. Though biblical in origin, this story maps perfectly onto the story of evolution as well.[48]

In the environment where we evolved, we drank when thirsty, ate when hungry, and worked when motivated. Our urges and what was urgent were the same. When we started to anticipate the future, to plan for it, we put ourselves out of step with our own temperament, and had to act not as nature intended.[49] We are all hardwired with a time horizon that is appropriate for a more ancient and uncertain world, a world where food quickly rots, weather suddenly shifts, and property rights have yet to be invented. The result is that we deal with long-term concerns and opportunities with a mind that is more naturally responsive to the present. With paradise lost and civilization found, we must forever struggle with procrastination.

Bottom line: procrastination is not our fault, but we have to deal with it nonetheless. We encounter procrastination across almost all of life's domains, from the boardroom to the bedroom, shifting from major to minor. Is it your home life, your finances, or your health that suffers the most from procrastination? Is it your e-mail or TV habit sucking away your productivity? Odds are, not only is the amount of your procrastination increasing, but so is the number of places you are doing it. But I am getting ahead of myself—that's the subject of the next chapter.

CHAPTER FOUR

ProcrastiNations

HOW MODERN LIFE ENSURES DISTRACTION

Over the bleached bones and jumbled residues of numerous civilizations are written the pathetic words: Too Late.

MARTIN LUTHER KING JR.

Our love affair with the present moment is at the root of procrastination. The fact that we tend to be more impetuous than reasonable is an evolutionary heirloom handed down through a thousand generations. But we can't blame our neurobiology entirely. Every distraction the modern world offers also exacerbates the mismatch between who we are and what we need to be. This chapter diagnoses the growing divergence between our plans and our impulses. To better write it, I reacquainted myself with an old distraction, purposefully re-infecting myself with what had afflicted me for so long as a student—video games. Gaming's capacity to absorb and dominate my attention is remarkable. There have been days, which stretched into nights, when I tore myself away from the screen only long enough to cram fast food into my mouth and to take care of my bodily needs. Eventually, I winnowed down every scheduled responsibility or meeting to its barest essentials in order to minimize the time it took away from gaming. My girlfriend viewed it as my mistress. My motto was Just One More Turn.

So, for the purposes of this book, I decided to explore Conquer Club, an online version of Risk©.[1] I hadn't played the board game since my college days, rolling dice with a few friends and beers, so I found its nostalgic aspect attractive. Also, Conquer Club's free version only allowed you to play four games at a time, so I believed it would be difficult for it to get out of control. Since you are playing against people around the world, moves are done at all hours and the game progresses at unexpected moments. Consequently, I found myself checking the website fairly often, even when it wasn't my turn. Suddenly there it was—procrastination, in the familiar form of gaming when I should be working. I could feel the hooks sinking deeper into me and I knew I was taking an extended walk along a sharp precipice—you know, the one you are going to slip over sooner or later (and secretly you are full of anticipation and excitement waiting for the stumble).

My, how quickly addictions reassert themselves. One Friday night, at the end of a long unrewarding work week, with sick kids, and after a trivial tiff with my wife, I felt life owed me more. Perhaps it was a bad idea to then upgrade to the premium version of Conquer Club and play twenty-five games simultaneously.* Periodic checks on the progress of the game became my life's punctuation marks, the periods capping off any of my tasks. At any break in the day's flow, I would peek

* It was *definitely* a bad idea.

into my games and see what battles had been fought in my absence or perhaps (joy, oh joy!) take my turn. Conquer Club continued to draw me back for weeks after that fateful Friday night. I checked on my games before and after the drives to work and home. It was the last thing I did before I went to sleep, the first thing I did when I woke up; and I dreamt about it in between. Oh, the sacrifices I make for science! But don't worry about me. Being both a victim and a detective of the fine art of distraction has its advantages. I know how to wind down this obsession, which I will do just after recapturing Kamchatka. While I wait for my turn, let's talk about why you, along with everyone you know, have likely experienced similar problems.

FULL-IMPULSE DRIVE

One of the elements that made me a slave to Conquer Club corresponds perfectly to the first and strongest findings from my research program: proximity to temptation is one of the deadliest determinants of procrastination.[2] Since every computer offers an opportunity to play, it is hard to keep temptation at bay. The second element is the virulence of the temptation; the more enticing the distraction, the less work we do. Conquer Club followed what is known as a "variable reinforcement schedule"—that is the reward (i.e., reinforcement) occurred at unpredictable times. For over fifty years, ever since B.F. Skinner and C.B. Fester's 1957 magnum opus *Schedules of Reinforcement,* we have known that such variable patterns of reinforcement are very addictive.[3] Skinner found that from pigeons to primates, we all work much harder for rewards when they are unpredictable but *instantaneous* when they arrive. You can see the power of variable reinforcement in gambling. Slot machines are fine-tuned for addiction because they have these schedules of winning hardwired into them. Every time a grandparent spends the grandkids' inheritance on these one-armed bandits, you can give a nod to the wonderful power of motivational psychology.[4] Unfortunately, as my Conquer Club example confirms, the Internet has given rise to a variety of similarly structured diversions. Paradoxically, while the Internet has made it easier for us to work, it has also created a series of behavioral traps that make it harder for us to work at all. In case it helps, I put this in graph form. It shows what comes between our wanting to accomplish a task and our ability to actually complete it.

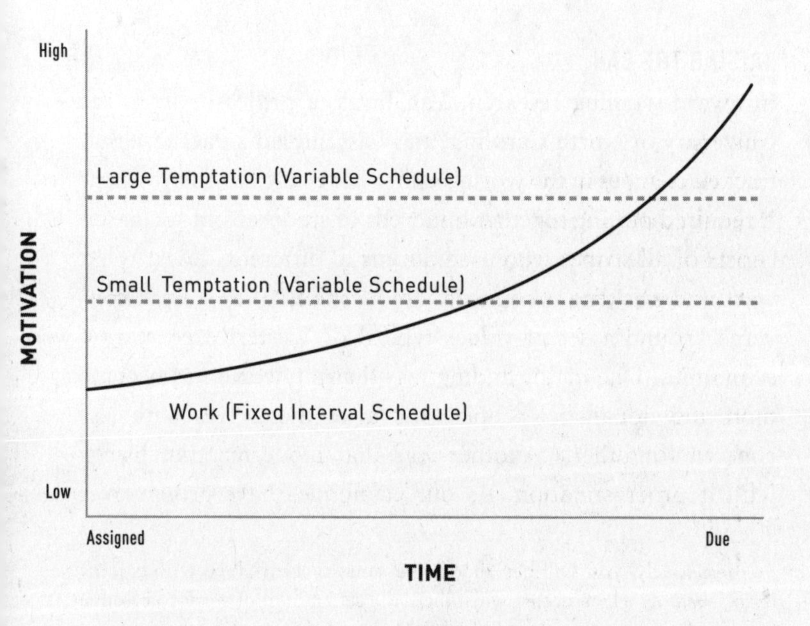

The graph's two horizontal dashed bars represent temptations, the lower bar being a small temptation (something nice) and the higher bar being a large one (something great). The solid line that eventually swoops up is the work curve, showing that, as we have seen, most of one's motivation is reserved until just before the deadline.[5] This is a *fixed interval schedule,* meaning that there is a fixed deadline before your work is assessed and you get "paid."* On the other hand, variable reinforcement schedules (the horizontal dashed bars representing small and large temptations) exert a constant state of motivation, typically much higher than fixed schedules. The motivation to play is always there and doesn't go away. The more attractive we make the temptation (making it large instead of small), the higher its bar moves and the longer it takes for the competing work line to become the dominant choice. So, we can see that when the allure of temptations rises, so does procrastination.

RAISING THE BAR

In award-winning research, Vas Taras, a professor from the University of North Carolina, and I assembled a database that tracked changes in the world's culture over the last forty years.[6] It required piecing together hundreds of studies from social scientists of all stripes who used dozens of different scales. What we found was that as countries "modernize," they start to converge around a set of values typical of Western free market economies. One major finding was that the world has become more individualistic: people look after themselves with less concern for others. Another was that modernization brings with it procrastination. As our economies have grown over

---

* Alternatively, you will get almost the same type of slope with a *fixed ratio schedule,* which occurs when there is a set amount of work to be done before reaping your reward. For example, piece-rate factory workers who get paid for every one hundred units produced tend to work a little harder as they approach that hundredth mark and then they take a breather. In the professional literature, this is known as "break and run," the pattern of taking a break after completing a work block before accelerating once again toward the next finish line.

the last few decades, we have experienced a fivefold increase in chronic procrastination. In the 1970s, 4 to 5 percent of people surveyed indicated that they considered procrastination a key personal characteristic. Today, that figure is between 20 and 25 percent, the logical consequence of filling our lives with ever more enticing temptations.

Consider how the world was transformed during the last century. In 1911, William Bagley, writing in *The Craftmanship of Teaching*, described the "hammock on the porch," the "fascinating novel," and the "happy company of friends" as the "seductive siren call of change and diversion, that evil spirit of procrastination!" Bagley's temptations, though real, were relatively pedestrian compared to what was to come. Also in 1911, the first film studio opened in Hollywood, and the next few decades saw the rise of multi-million-dollar productions and multi-millionaire movie stars, along with their scandals; both Charlie Chaplin and Errol Flynn—the comic tramp and the romantic swashbuckler—seemed to like their women a little bit younger than the law allowed. The spectacle of Cecil B. DeMille's *The Ten Commandments* hooked viewing audiences, and by the 1930s, the popular press was referring to movies as a common form of procrastination.[7] Still, you had to leave your home or office to see the silver screen. But not for long. The end of the Second World War coincided with the development of television, and the number of Americans with TVs leaped from 9 percent to 65 percent between 1950 and 1955. During popular show times, streets would empty and stores would shut so that everyone could tune into episodes of *I Love Lucy*. By 1962, with television sets now in 90 percent of American homes, *Popular Science* published a book digest, "How to Gain an Extra Hour Every Day," connecting television watching to procrastination.[8]

In the mid-1970s, a new temptation emerged on the scene. I was eight years old when Pong, the first successful video game, was introduced to our household. My father hooked up one end of the game box to our black-and-white TV and the other end to two "paddles" that were nothing more than knobs on cords. You turned your knob and a small bar moved vertically on either the left or right side of the screen, depending on which paddle you had. If a moving electronic ball hit your paddle, it was deflected back to the other side of the screen for your opponent to return. That's it, but it was magic and I loved it. Sure enough, by 1983, psychology texts were reporting video gaming in the list of typical procrastination behaviors.[9]

Having looked at some historical baselines for comparison, you should be able to see why procrastination has risen to today's levels. While the pleasure derived from working has remained fairly constant over the decades, the power of distractions only seems to increase. The temptation bar in Skinner's graph is raised ever higher, while the work curve remains the same. Let's reconsider today's video games, which make Pong laughable by comparison. Unfathomably more advanced, these games are the product of untold millions of programming hours, and tax the capacity of even the most advanced computer systems. They beat hands-down anything that Bagley wrote about. Many people play anywhere, anytime—it is not uncommon for students to engage in head-to-head online games during university lectures.[10] Furthermore, as good as these games are, they are getting better. With each evolving iteration of Grand Theft Auto, Guitar Hero, or World of Warcraft, choosing *not* to procrastinate becomes harder. The graphics, the story, the action, the console—all of them advance. In the battle for your attention, it is as if work is still fighting with bows and arrows while gaming has upgraded to auto-cannons, sniper rifles, and grenade launchers. Consequently, it is becoming increasingly common for people of all ages to become consumed by games, and intervention centers are proliferating to treat video game addiction. In Korea, for example, about 10 percent of young people show advanced signs of addiction, developing up to seventeen-hour-a-day habits. In response, the government has sponsored 240 counselling centers or hospital programs. There are even resources for particular games, such as www.WoWdetox.com, which is dedicated to World of Warcraft players as well as their spouses, typically called Warcraft widows.

What is more shocking is that there are worse creatures than video games for inciting procrastination. The worst isn't getting a bite to eat or napping, though they remain popular choices. The king of distraction—and there is only one—is television.[11] Since its halcyon years in the 1950s, television has continued to perfect itself, gaining all the features it needs to win the competition for our time. The magic of the remote allows us to change channels without moving. The advent of cable and satellite has ensured that there is always at least one available channel that reliably caters to our tastes. And with multiple television sets throughout the house—more TVs than people according to Nielsen Media Research—we can watch our shows anywhere we like. If our interest in a particular program flags even momentarily—zappp!—we are off to other worlds in this 500-channel universe. So attractive is television that we are often guilty of over-consumption, feeling TV'd out and wishing at day's end that we'd watched a little less.[12]

Most Americans spend about half of their leisure hours in television's glow. Other nations aren't far behind. According to the latest national census data, Americans watch an average of 4.7 hours per day, beating out Canadians, who watch 3.3 hours per day. The average Thai spends 2.9 hours in front of the tube; a Brit, 2.6 hours, and a Finn, 2.1 hours. Reading, for comparison's sake, clocks in at an international average of 24 minutes a day. This means, of course, that you have likely been plugging through this book for about three months now.

Worse still, as with video games, TV is getting more and more attractive. Not only is the hardware becoming sleeker, and higher-tech, but your options of what to watch are also stepping up. Season box sets are commonplace, as are digital video recorders (DVRs). These DVRs allow you to record multiple programs simultaneously, store hundreds of hours, keep track of what you have watched, and help you find desirable episodes. Watching scheduled TV seems primitive today. The future promises even more. As television continues to evolve, viewing options become almost endless. For example, the technology already exists to download any movie in less than a second. When a fraction of that power becomes available to the ordinary household, we can expect our TV viewing to rise correspondingly. Any movie, any show, any clip, can be seen by anyone, almost anywhere, all at an insanely crisp resolution. Inevitably, as television pumps up, it muscles out the rest of life. It is already happening. For every country with data, the amount of television watching has increased. In just eight years, from 2000 to 2008, TV watching in the United States went from 4.1 to 4.7 hours, a 15 percent increase. Since time is finite, everything else must suffer—and suffer it does.[13] It's not just chores that we put off in favor of television—it's eating with the family or connecting with friends.

I have painted a pretty bleak picture, but it could be worse. Actually, the worst is still ahead. It doesn't quite take up as much time as television yet, but it shows a lot more potential. It is the Internet, which has all the allure of video games, television, and more on a single platform. About 80 percent of students already report that Internet activities are particularly problematic for them.[14] No wonder. There are websites and blogs that cater to every fetish and interest, videos to download, music to acquire, and text-messages to respond to. An interesting twist in the Internet procrastination saga are the hundreds of social networking sites, like Facebook, Bebo, MySpace, and Twitter. Less than a year from its inception in early 2004, Facebook was reported in *The New York Times* as a key enabler of dawdling, with students hitting the refresh key on their screens hundreds of times a day to check for updates.[15] This habit is not dissimilar to Skinner's variable reinforcement research on rats and pigeons, who also pressed their keys hundreds of times for an eventual but unpredictable reward pellet. Since researching Facebook in person would likely endanger the hours I needed to write this book, I decided to locate an expert, someone already intimately entangled and familiar with the site's details. As if to underscore Facebook's prevalence in the university population, it took me less than five minutes to track down such an authority, specifically a graduate student. She "Facebooks" about ninety minutes a day and has even set up a Dogbook page for her pug, Schmeebs. I sat down beside her computer so she could give me the tour:

- "The first thing," she explains, "is that this gives you a way to connect to all your friends and controls what type of contact you want from them. For example, I am interested in other people's pictures, so this section provides previews of their photographs from their Facebook pages."
- o I comment that it seems to be a large section of the screen.
- "Oh, well I have quite a few Facebook friends."
- o How many?
- "Let me see," she responds. "Here it is, 603."
- o That does seem like a lot. Do you really have 603 friends?
- "No, no, a lot of them are just acquaintances and you can treat them differently. You control security and access to what you

get from them and what they can post on your website. There is a wall here, where my friends can post comments, see."

- I see.
- "Some get priority. My friend Jen is one of three people who I get text messages from every time she updates her Facebook site . . ."
- Which is?
- ". . . about twice a day."
- How long do you wait until checking what the update is about?
- "Well, the text message is usually incomplete, so I have to go online to read the whole thing."
- So, immediately.
- "Exactly. Right away."
- Even if you're at a movie, at dinner, or with family?
- "Of course," she responds, "though if it is with family, I will sneak off first."
- Very polite of you. What else can you do?
- "So, so much. You can 'poke' people just to tell them you are thinking of them, you can give them virtual gifts . . ."
- Why?
- "Because you can. Some gifts are free or corporate-sponsored, others you have to pay for. Here is a bunch of booze-related ones. I am not sure why I have these. You can also use Facebook to send invitations to events."
- So this enables you to meet and interact with people more?
- "No, not really. It has replaced a lot of my socializing.
- But, I do feel I have gotten to know a lot of my closer friends better. You post funny quotes you have heard, videos of trips, whatever tidbits interest you. Oh look! Chelsea's dog added Schmeebs as a friend!"

My expert then leads me to the myriad procrastination-dedicated Facebook groups, such as: "AP-Advanced procrastination" (over 18,000 members); "I'm majoring in Napping and Facebook with a minor in procrastination" (over 30,000 members); "I was doing homework and then I ended up on Facebook" (over 900,000 members). There are also over 600,000 members of a Fan Page dedicated solely to "Procrastination" itself. Joining these groups makes a statement about your identity, and provides a wide assortment of suggested diversions, as well as an opportunity to talk about them. Ironically, a recurring discussion among members was focused on how they could limit or quit Facebook itself (e.g., "let parents change your password and only tell you after exams"). Not to be perceived as a Facebook prude, I agree that the site is attractive and intriguing, and that it has useful applications, especially networking. Napoleon Hill, an achievement guru of the last century, considered networking a key element of success. On the other hand, Facebook is a tremendous distraction and it is this and not the networking itself that dominates. A true sign of the addictive aspects of Facebook is that half of those who quit reactivate their accounts.[16] They can't keep away.

HOW WE GOT HERE

The rise of procrastination is hard to avoid, given its deep roots in our brain's neurobiology. The limbic system focuses on the now while the prefrontal cortex deals with longer-term concerns. In other words, when building a fire, the limbic system is eyeing that can of gasoline while the prefrontal cortex says branches and logs would provide slow, steady heat. The first wants the immediate million-dollar check, while the second favors a weekly five grand for life. Though both the limbic system and the prefrontal cortex come together to reach a final decision, their duet ensures the rise of procrastination. Here's an example to show you what I mean.

Let's take two snack food vendors, *Nutrity Nuggets* and *Tasty Tempts. Nutrity Nuggets* provides healthy fare, which appeals to our long-term and more abstract goals of a slender waist and improved physical well-being. This is brain food . . . well, at least its prefrontal portion. *Tasty Tempts* provides sugar and fat in a dozen deep-fried combinations, immediately delicious and sure to tickle the fancy of our limbic system. Set up across from each other at the mall, which of these two stores will sell more snacks? You don't need a degree in marketing to conclude that sugar's moment on the lips wins out over its lifetime on the hips. *Nutrity Nuggets* will be most people's choice of tomorrow, what they *intend* to eat, while *Tasty Tempts* is the choice for today, what they *are* eating. Furthermore, as sure as the high price of movie theater popcorn, *Tasty Tempts* delivers a much larger profit margin because impulsive purchasing curtails comparison shopping. *Nutrity Nuggets* may well go under while *Tasty Tempts* becomes an international franchise. Businesses respond to our dominant desires, so there is no coercion or conspiracy here, just the invisible hand of the market building a limbic system wonderland. With the ubiquitous overemphasis on the immediate and the material, on the instant and the consumable, people are seduced into putting off long-term but ultimately more satisfying goals involving career achievement, volunteering in the community, raising a family or following a spiritual path. Materialism and consumerism are merely emergent properties of our neurobiology given free rein in a free market.

The process of seduction all starts with the sophisticated science of market research. I should know; I met my wife, Julie, while she was getting an advanced degree in this area. Market research has many different applications, some of which are benevolent. For example, my wife's university adviser conducted research on how to craft warning labels on cigarettes so that people *don't* buy a pack. But, like most of life, market research usually concentrates on where the money is, and from children's television to political parties, marketers are adjusting products to our tastes or even creating desire for them. In doing so, they invariably appeal to our limbic system by creating temptations. The food industry, in particular, has used market research to the hilt, finding out what is most delicious to consumers and how best to package it. In his book *The End of Overeating,* Dr. David Kessler, a commissioner of the U.S. Food and Drug Administration under two presidents and the former dean of the Yale School of Medicine, investigates the incredible vigor and purpose that food producers commit to getting us to eat more of their cheap and nutritionally challenged products.[17] The amount of engineering that goes into creating visually appealing, flavorful food that scrumptiously crunches, melts, or rolls in your mouth is astounding, rivaling only by that put into designing your flat screen TV or Blu-Ray player. With exquisite fine-tuning of the proportions of sugar, fat, and salt in a recipe, for example, dishes can be created that provide no satiation—we always have room for one more bite.

Once a product has been devised, the emphasis on the limbic system continues during its presentation. Advertisements, which comprise about 1 to 2 percent of most economies, typically accentuate the most concrete and salient aspects of any merchandise.[18] When you next walk through a grocery store, note how prominently the look or taste of any product is displayed; compare that to the extra effort required to find its nutritional content or its relative cost, both of which would appeal more to your prefrontal cortex. Finally, the virulence of temptations is truly maximized if desired products are immediately accessible;

availability encourages the impulse purchase.[19] Since this principle of instant availability considerably strengthens the role of the limbic system in our decision making, we see a lot of it. "Buy-now, pay-later" sales strategies stress the present moment, and sales gurus like Zig Ziglar emphasize: "If the decision is yes, then you . . . could be enjoying the benefits NOW!"[20] As David Mesla, a weight management scientist at the Unilever Health Institute, notes: "Every single day and every single place you go, those foods are there, those foods are readily available for you to engage in. There is constant, constant opportunity."[21] Universal proximity is exactly the goal—to shave enough seconds off the mechanisms of delivery that all products can be purchased as impulsively as the candy by the checkout counter. Once this happens, the world becomes an inescapable cage of temptation and if your willpower ever lapses, even for just a second, that's all the time they need to get you. But wait, there's more.

Aside from making their products and presentations more alluring to your limbic system, marketers also make concerted efforts to push that pesky prefrontal cortex aside. Habits and rituals, in particular, bypass the prefrontal cortex during decision making, and so great efforts are made to cultivate them in consumers.[22] Many of our purchases are triggered rather than chosen, just as the addictive aroma of KFC's eleven special herbs and spices is designed to create a sudden craving for deep-fried chicken. Regardless of our original intentions, once cued, our actions can be "emotionally hijacked" and we end up automatically eating fast food or ordering that specialty coffee.[23] We are all vulnerable to this, me included; I am guilty of consistently buying overpriced movie theater popcorn, which science confirms is less of a conscious choice than a ritualistic act cued by entry to the theater.[24] Research on exactly how vulnerable we are is the specialty of Brian Wansink, a professor of consumer behavior at Cornell University. His studies have established that our eating choices are indeed mostly routines that are weakly based on hunger and more strongly based on context, such as the size of plate or portion, or how visible the treat is. For a study that won him the Ig Nobel Prize, he infamously devised a bottomless soup bowl that surreptitiously refilled itself as people ate from it.[25] Though they reported not feeling any fuller than those who had a single bowl of soup, those who ate from the ever-full bowl consumed almost twice as much—76 percent more. Such habits now make up about 45 percent of our daily actions, and increasing this percentage by providing easy options and clear cues is big business.[26]

The extent to which we can become wedded and welded to our habits is best revealed by the way we rely on our smartphones, like Apple's iPhone or Research in Motion's BlackBerry. People text everywhere, even while driving. This is a salient example of a pleasurable impulse overriding good judgment, for, as common sense concludes, and my own as well as others' research shows, any cell phone use while driving (hands free or handheld) slows one's reaction time dangerously.[27] Such are the addictive qualities of smartphones that the World College Dictionary voted "CrackBerry" as the word of the year in 2006. They are so embedded in people's lives that at times our brains, in a testament to their neural plasticity, adopt the devices as part of our bodies. When it isn't around, people feel something like phantom limb syndrome (sometimes termed "fauxcellarm"). Others report the more familiar problem of repetitive motion disorder, like "BlackBerry Thumb," which is recognized by the American Physical Therapy Association as an official workplace injury. And what are people frenetically doing with their smartphones so much as to wear out their joints and ligaments? With tens of thousands of applications available, the company comScore undertook to categorize the top twenty-five applications downloaded into the iPhone. The only one that wasn't entertainment, a game, or a social networking site was Flashlight—a utility that turns your iPhone into a light source.[28]

SUPPORTING VOICES

So here's the situation. Procrastination isn't just battling a hundred million years of evolution. It is battling a hundred million years of evolution that are being actively exploited at every turn by the very fabric of our society. In 1958, Aldous Huxley, in his book *Brave New World Revisited,* warned: "All the resources of psychology and the social sciences are mobilized" with the aim of controlling people by finding "the best ways to take advantage of their ignorance and to exploit their irrationality."[29] In 1985, with the rise of the video game, the influential cultural critic Neil Postman specifically built on Huxley's work in *Amusing Ourselves to Death,* where he points out that "the rationalists who are ever on the alert to oppose tyranny 'failed to take into account man's almost infinite appetite for distractions.'"[30] And we have Avner Offer, professor of economic history at Oxford University, reviewing in 2006 how current Internet consumption contributes to many of the world's ills, in *The Challenge of Affluence.*[31] In sum, the free market is geared toward providing increasingly irresistible temptations that distract us from our greater goals.*

Are we right, Huxley, Postman, Offer, and I? Well, look around you. Exactly how many recreational or entertainment pursuits do you have readily available? A typical household can hold

* Similarly, and at about the same time, the psychologist Stuart Vyse reports in *Going Broke: Why Americans Can't Hold On to Their Money,* "Any time the urge strikes, we now have the capability to act on it impulsively, and that creates a much greater challenge for us than was ever the case before. It's only natural that we are having trouble with debt."

hundreds, from widescreen TVs to Internet portals. Never before in our history have there been as many temptations, as succulently devised, as readily available, and as adeptly marketed. Adam and Eve only had to deal with a juicy apple purveyed by a serpent. Nowadays, our apple is caramel coated and chocolate dipped, marketed with a multi-million dollar advertising campaign in a blitz of commercials, pop-ups, and inserts.[32] Inevitably, as our lives drown in these diversions, our procrastination is on the rise.

LOOKING FORWARD

There is no turning our backs on modern life. The free market, in one form or another, will continue and the pace of invention will only accelerate. We will benefit from many of these innovations, but not all. The exploitation of the limbic system is baked into capitalism and you can't stop it without making the entire wonderful wealth-generating machinery grind to a halt. Someone will always create a product that provides short-term pleasure along with considerable but deferred pain simply because we will buy it. Consequently, dealing with constant temptation and its potential for creating procrastination is and will continue to be part of living in this world. Good thing you are reading this book, then. Learning better ways to cope with temptation and procrastination is what we will be doing together in later chapters; we will make the Procrastination Equation work for us, one variable at a time. But first, let's whet your appetite by acknowledging what procrastination costs you and society at large. A single incident of procrastination can be petty, but once you see the ultimate toll, I think you will find that it is an opponent worth fighting. I surveyed four thousand people to find out where they procrastinated most. The next chapter reveals what they told me and the personal price they paid for putting things off.

CHAPTER FIVE

The Personal Price of Procrastination

WHAT WE MISS, WHAT WE LOSE, AND WHAT WE SUFFER

We have left undone those things which we ought to have done; and we have done those things which we ought not to have done.

THE BOOK OF COMMON PRAYER

For enduring fame and sheer depth of procrastination, Samuel Taylor Coleridge stands alone. One of the great poets of the nineteenth-century romantic era, Coleridge might have been its greatest, but that title is more often given to his one-time and more diligent friend, William Wordsworth. Coleridge's tragic weakness was procrastination. He put off his work and his obligations, at times for decades. The poems for which he is best remembered, and which are still regularly studied in English literature classes, all show traces of procrastination. *Kubla Khan* and *Christabel* were both eventually published as fragments—unfinished works—nearly twenty years after he began them, and *The Rime of the Ancient Mariner,* though completed, was five years late to press.

Everyone—his family, friends and even he himself—recognized Coleridge's procrastination. His nephew and editor, Henry, wrote that his uncle was "the victim of a procrastinating habit," and Coleridge himself describes his procrastination as "a deep and wide disease in my moral Nature . . . Love of Liberty, Pleasure of Spontaneity, these all express, not explain, the fact." However, it was his close friend Thomas de Quincey who provided the best account, having shared with Coleridge not only a proclivity to procrastinate but also a severe drug addiction—Quincey's autobiography is aptly titled *Confessions of an English Opium-Eater.* As Quincey wrote:

> I now gathered that procrastination in excess was, or had become, a marked feature in Coleridge's daily life. Nobody who knew him ever thought of depending on any appointment he might make. Spite of his uniformly honourable intentions, nobody attached any weight to his assurances *in re future* [in regard to the future]. Those who asked him to dinner, or any other party, as a matter of course sent a carriage for him, and went personally or by proxy to fetch him; and as to letters, unless the address was in some female hand that commanded his affectionate esteem, he tossed them all into one general *dead-letter bureau,* and rarely, I believe, opened them at all.

Coleridge's excuses for lateness have themselves become legendary. His correspondences consist frequently of apologies, at times even an extended run of them; witness his letters to a Mr. Cottle, a publisher who bought the copyright to a book of his poems—sadly, in advance. Deserving special mention is the "Person from Porlock," who Coleridge claimed irrevocably interrupted his recollection of the opium-induced dream that served as the basis of his poem *Kubla Khan.* The poem runs only 54 lines, instead of the intended 200 to 300. As judged by Robert Pinsky, an American poet of our time, the "Person from Porlock" is the most famous of fibs from a long line of writers who are "better at making excuses or self-indictments than at getting things written."

But what did this procrastination reap for Coleridge? As Molly Lefebure describes his situation in her book *A Bondage of Opium,* "his existence became a never-ending squalor of procrastination, excuses, lies, debts, degradation, failure." Financial problems pervaded his life, and most of his projects, despite elaborate planning, were barely begun or finished. His health was terrible, exacerbated by his opium addiction, for which he delayed medical treatment for an entire decade. His enjoyment of work dissolved in the stress of unmet deadlines—"My happiest moments for composition are broken in upon by the reflection that I must make haste." He lost rare friends, such as Wordsworth, and his marriage ended in separation because of it.

Coleridge's woes clearly illustrate that procrastination is capable of damaging all aspects of our lives. However, only the most confirmed of procrastinators will experience anything that approaches Coleridge's sad life. Most of us procrastinate significantly in only a few of life's domains. To find out about the procrastination habits of ordinary people living in our own time, I put up a survey on my website (www.procrastinationequation.com) and four thousand people answered it. I asked them to tell me how much they procrastinated in each of twelve major life domains and to rank what was most problematic for them.[1] The table on the next page reflects the results. The first column is the domain, and the second column records people's average level of procrastination, with a score of 2 indicating *seldom,* 3 indicating *sometimes,* and 4 indicating *often.* The third column indicates the percentage of people who chose that life domain as a "top three problem."

TWELVE MAJOR LIFE DOMAINS	AVERAGE PROCRASTINATION (1 TO 5)	A TOP THREE PROBLEM
1. HEALTH: exercise, diet, avoiding or treating illness (e.g., "Procrastinating about going to the gym, about going on a diet")	3.4	42.2%
2. CAREER: jobs, employment, earning a living (e.g., "Procrastinating about getting a better job, about getting a raise")	3.3	56.8%
3. EDUCATION: school, studying, getting good grades (e.g., "Procrastinating about studying for an exam, about getting a degree")	3.3	32.9%
4. COMMUNITY: volunteer work, political activism (e.g., "Procrastinating about helping others, about volunteering your time")	3.2	12.1%
5. ROMANCE: love, sex, dating, marriage (e.g., "Procrastinating about asking someone out, about ending a relationship")	3.0	24.0%
6. FINANCE: decisions about money (e.g., "Procrastinating about paying your bills, about saving for retirement or a big purchase")	2.9	35.9%
7. SELF: improving oneself in terms of abilities, attitudes, behaviors (e.g., "Procrastinating about reading a self-help book, about changing who you are")	2.9	29.6%
8. FRIENDS: interactions with close others (e.g., "Procrastinating about spending more time with your friends, about having them over")	2.9	23.5%
9. FAMILY: interactions with parents and siblings (e.g., "Procrastinating about talking more to your mom, about having dinner with your parents")	2.7	18.9%
10. LEISURE: sports, recreation, hobbies (e.g., "Procrastinating about joining a sports team, about going on a trip")	2.7	11.4%
11. SPIRITUALITY: religion, philosophy, the meaning of life (e.g., "Procrastinating about examining your life, about going to church/temple/mosque")	2.5	8.5%
12. PARENTING: interactions with offspring (e.g., "Procrastinating about spending time with the kids, about having a family vacation")	2.3	4.1%

As you look at this table, pay attention to the domains where the numbers in the last two columns are both high: these are the trouble spots.

Procrastination causes us grief at school, at work, and in our private lives, particularly in relation to health. Eighty-nine percent of people believe they have major problems in *at least* one of these three life domains alone, with 9 percent approaching Coleridge's levels by reporting all three. There is also a pattern to people's procrastination; most of these life domains cluster or hang together in groups. For example, many of the people who reported procrastinating about their financial situation also reported putting off education and other activities that might improve their careers (domains 2, 3, and 6). If you are suffering in one of these three areas, you're likely to feel that the other two are also problematic. This "Success" cluster of concerns is overall the most prevalent in terms of procrastination.

A second cluster focuses on "Self-Development," as people who put off their health issues (domain 1) also tend to put off spiritual quests, leisure activities, and self-improvement programs (domains 7, 10, and 11). This is the broadest cluster, as it is also connected to your social life, being part of your community, or pursuing a romance (domains 4 and 5). A final cluster could be labeled "Intimacy" by virtue of grouping together close friends, family, and parenting (domains 8, 9, and 12). This is the least problematic of the lot, especially in regard to parenting. Happily, very few respondents report putting off raising their kids—there is an immediacy to caring for children that trumps everything else.

Whether your procrastination lies in the Success, Self-Development, or Intimacy cluster determines the price you pay for procrastination, as these three areas translate into three major costs: your Wealth, Health, and Happiness. Naturally, those who put off the Success cluster and its career or financial aspects will be less wealthy. Those who procrastinate on Self-Development will experience poorer health, both of body and spirit. And though happiness is affected by the previous two clusters, Success and Self-Development, it has the strongest ties to Intimacy. In a different meta-analysis that I conducted, based on close to twelve hundred studies, I established that the biggest predictors of happiness are traits leading to fulfilling interpersonal relationships; great wealth and good health mean less without someone to share them with.[2] Wherever your procrastination lies, the more you do it, the greater the cost. Just take a look.

FINANCIAL PROCRASTINATION

The most common excuse I hear from people who procrastinate at work is that they are more creative under pressure. I can see how it might appear this way. If all your work occurs just before a deadline, that is when all your insights will happen. Unfortunately, these insights will be relatively feeble and few compared to the insights of those who got an earlier start, since under tight timelines and high pressure people's creativity universally crumbles.[3] The bleary-eyed 3:00 a.m. crowd scrambling to finish a project will usually come up with routine, unremarkable solutions. Innovative ideas are typically built on the bedrock of preparation, which includes a laborious mastery of your topic area followed by a lengthy incubation period.

Other procrastinators try to justify their delays by indicating that they work most efficiently closer to the deadline. This time, they are partly right. You do have more motivation just before the clock strikes twelve and you cross the deadline. But what the procrastinator is creatively arguing here is not whether we work hardest at the eleventh hour (which we do) but that working *earlier* actually harms our performance. In other words, this procrastinator says that working today *and* tomorrow is worse than working only tomorrow—a clumsy lie.

No matter what index of success we examine, procrastinators tend to perform worse than non-procrastinators. There is some variation based on whether we look at education, career, or income, but not in a way you are going to like: as procrastination moves from school to job to measures of overall wealth, the worse its effects. The results (which I summarize in my article "The Nature of Procrastination") look like this.[4] For high school and college students, only about 40 percent of procrastinators have above-average grades, while 60 percent are below. If you are one of the lucky 40 percent, you should recognize that though procrastination is a handicap, you are probably compensating for it with other attributes, like a brilliant mind. Don't fall into the trap of thinking that this vice is actually helping you out. Not that I should judge . . . I put off studying for too many finals and tried to erase my late start by doing a series of all-nighters, a strategy that ended with me dozing peacefully through the last half of a French exam. The consequence of that impromptu nap has me dreading foreign languages to this day. The funny thing is that almost everyone else has a similar story to share.

Students spend roughly a third of their waking hours on diversions they themselves describe as procrastination.[5] On average, students engage in over eight hours of leisure activities on the two days prior to exams,[6] and their inability to effectively manage their time is a self-reported top concern as well as a reason for dropping a course.[7] Worse, this trend doesn't abate when the stakes get raised. It is a major reason why most potential PhDs leave school before graduation, and the three little letters they get after their name are ABD (all but dissertation).[8] ABDs are so common that cartoonist Jorge Cham, for example, makes a living by writing *PhD Comics,* a strip dedicated to chronicling PhD students' procrastination. Incredibly, after graduate students have gotten into a competitive academic program, done all their course work, perhaps even gathered their dissertation data, and need only to write it up and defend their thesis, at least half never complete the process despite the immense investment of time and the significant rewards for completion (on average, a 30 percent increase in salary).[9] Procrastination is the primary culprit.

Moving on to the field of career success, we find that the impact of procrastination intensifies a little bit. As judged by their peers, 63 percent of procrastinators are in the below-average, unsuccessful group. From the get-go, procrastinators have trouble getting going, and they put off the job hunt. When unemployed, they stay that way for longer.[10] Once employed, most will find that work life is less forgiving than college or high school. The stakes are higher, so it is harder to get your boss or your clients to give extensions. For example, Michael Mocniak, general counsel at Calgon Carbon, got fired from his job for putting off processing his invoices—$1.4 million worth.[11] Furthermore, on-the-job projects can be larger and much harder to complete at the last minute; work is less predictable, so you could find the final hours before a deadline suddenly double-booked. Still, there is that 37 percent of procrastinators whose wealth is at least above average despite their character flaw. If over the breakfast table you call the CEO mom and the Chairman of the Board dad, you are likely going to be wealthy no matter what your personal vices are. Other chronic procrastinators might end up in a career—and there are a few—where it is very difficult to procrastinate—careers with built-in daily goals, like sales or journalism. With everything due today, the leeway to procrastinate is exceedingly slim.

Finally, when we talk about overall financial success, the procrastination numbers again step up. By their own admission, only 29 percent of procrastinators consider themselves successful, with the remainder describing themselves as below average. The reasons for this are legion: procrastination's harmful touch extends into dozens of nooks and crannies that affect your bank balance.[12] For example, the U.S. government gets at least an extra $500 million each year thanks to tax procrastination. A typical procrastinator's mistake is failing to sign the forms in the last-minute rush, making them invalid and subject to a late penalty.[13] But procrastination hits our savings and spending in many other ways.

Savings speak to what Albert Einstein called the eighth wonder of the world—compound interest. The money you save not only earns interest, but the interest earns interest, like your children having grandchildren. Such is its power that if you put aside $5,000 each year between the ages of 20 and 30, you would retire richer than if you started putting that five grand aside *every* year from the age of 30 on. Alternatively, consider the native Indians who sold Manhattan Island for about $16 worth of beads. Had they taken the money and invested it, with compound interest they could pretty much have bought back the entire island today and everything on it—from the Christmas trimmings at Rockefeller Center to the boardroom leather chairs at Trump Tower.[14] Too bad procrastinators rarely act on their intention to sock money away for retirement or even a rainy day. If they were characters in one of Aesop's cautionary tales, they'd play the grasshopper instead of the ant. All that compound interest, all those potential investment dividends, lost and almost impossible to regain. As a paper in *The Financial*

Services Review concludes, "We find the levels of contributions required for individuals who start saving late are so high it is questionable whether they are affordable for anyone not on a high income."[15]

What's more, procrastinators tend to be credit card revolvers; that is, they have hefty rotating unpaid balances on their monthly statements. When you combine all the credit cards in a household, the total often exceeds $10,000.[16] Thanks to something called "universal default," most procrastinators are probably being charged the maximum rate for those balances, close to 29 percent per year, or 32 percent if it's on your Sears card and up to 113 percent if you live in Mexico. Universal default means that if you are late paying one bill, such as the phone or electric bill, the credit card company can jack up their rates. Make one mistake, anywhere, and they've got you (goodbye 0 percent introductory annual rate). Here compound interest again rears its head, but this time it's ugly. How are the credit card companies making record profits? At the dallying hands of procrastinators.[17] They affectionately call revolvers the "sweet spot" of their industry.[18]

Of all the scientific studies that show how procrastination is dangerous for your financial health, here is one I found particularly revealing. It deals with MBA students, so I've nicknamed them "Leaders of Tomorrow." The study demonstrates the self-defeating choices of procrastinators by examining University of Chicago MBAs, who often pride themselves on being cutthroat competitors.[19] After playing a series of games in which they could win up to $300, study subjects were given a choice about how they could receive their winnings. They could either get a check now or wait two weeks and get an even larger sum. Here is why procrastinators tend to be poorer: even though most of them demanded to be paid now, they didn't cash their checks until, on average, four weeks later. In other words, it took them *twice* as long to get to the bank as they would have had to wait for the larger reward. This brutal mixture of procrastination and impatience is common: two-thirds of the students wanted their money up front.

If none of this has resonated with you yet, I will add one final example, *the* final example: your last will and testament. Way back in 1848, Lewis Judson noted that procrastinators not only borrow excessively but they put off their estate planning too: "Most men postpone making their wills until on a sick bed, and often then, until too weak to make them clearly and the lawyers take more of the estate than the heirs."[20] In the ensuing one and a half centuries, nothing much has changed; right now, I bet your will is almost certainly either out of date or completely undone.* Though you will be dead when the ramifications of this particular procrastination play out, it is probably the ugliest possible legacy to leave for your friends and family.[21] Dying intestate—without a legal will—is common, happening to around three quarters of the population. George Gershwin (American composer), Richie Valens (rock pioneer), Howard Hughes (reclusive billionaire), Keith Moon (drummer for the Who), and Barry White (the smooth bass soul singer) all died intestate. Abraham Lincoln and Martin Luther King, Jr., despite both rallying against procrastination themselves and receiving a stream of death threats, died intestate too.† What dying intestate entails depends on the jurisdiction where you live. Your

* Not me, but only because I wrote this book. With two kids and a wife, I completed my will within a few days of writing this sentence.

† Abraham Lincoln: "The leading rule for the lawyer, as for the man of every other calling, is diligence. Leave nothing for to-morrow which can be done to-day."

Martin Luther King: "How soon not now, becomes never."

whole estate may go to the government, to a hated sibling, or perhaps to your ex-spouse from whom you are separated but not yet divorced (and his or her new partner). Nothing may go to your soul mate, your best friend, or your favorite charity, and all your family heirlooms could be sold at bargain prices. The law favors descendants over ancestors so, if you do have children, they could get your estate all at once at the supremely responsible age of eighteen with no strings attached.

MEDICAL PROCRASTINATION

Despite being an important medical procedure, the thought of a colonoscopy makes most people squeamish. Even a description, which I'm about to provide, often provokes discomfort. The first step in a colonoscopy is to clean out your innards as much as possible. Typically, this involves drinking a gallon of a very strong laxative until what you pass resembles what you put in. You may also need an enema, which requires you to take in another few quarts, except from the other end. After this cleaning, upside and down, you are ready for the physician. You will go to the hospital, put on a gown, and then be sedated. You certainly don't want to tense up when asked to lie on your left side and then receive a half-inch colonoscope through your rectum. A little air is usually injected to help inflate your bowel and allow a good look around. The doctor looks through the colonoscope and into you for about thirty minutes and your buttocks will feel greasy for a while afterward, but that's about it.

You should start getting these colonoscopies pretty regularly after the age of fifty, if not earlier, but a surprising number of people put it off, including oncologists. Even my father-in-law, who ran a large health sciences center and should have known better, unduly delayed his. It does sound unpleasant, but the downside of delaying a colonoscopy is potential death from colorectal cancer, the second most deadly form of cancer, right behind lung cancer. But unlike lung cancer, colorectal cancer is very treatable and preventable if you catch it early. It comes in stages, from 0 to 4, the survival rate plummeting with each successive stage. The number one reason for failing to get screened is procrastination. Putting off a recommended colonoscopy because of fear, discomfort, or embarrassment is a widespread problem even among the most capable. Katie Couric, while she was co-anchor of the *Today* show, lost her husband to it. My father lost his second wife. By the time she finally went to see her doctor, a colonoscopy wasn't needed because you could already feel the cancer through the wall of her stomach. After seeing her vibrancy slowly fade away in my father's care over her last year, I can confidently say that this is as serious and tragic as it gets. However, the story of colonoscopies is not unusual for medicine. For many diseases, infections, growths, and general ailments, early detection and treatment is always better, and yet people consistently delay. Given this lead up, I'm sure you won't be surprised to learn that procrastinators tend to be among the least healthy of people.

To rub salt into the wound, not only are procrastinators less likely to pursue treatments but they are more likely to indulge in the very behaviors that create the need for treatments in the first place. Procrastinators are health risks because their impulsive nature makes them susceptible to vices, attracting them to short-term pleasures despite their long-term pains. On the other hand, they are less predisposed to virtues—that is, short-term pains with long-term rewards. For example, do you floss? Though you know you should and often plan to, if you are a procrastinator, you very likely don't.[22] Exploring the effects of this oversight, I asked my dentist about the worst case he had seen. He recalled one patient with more tartar than tooth, tartar so thick that it formed a solid wall, obscuring the teeth. He offered to show me a picture; I wisely declined. Here are a few other misbehaviors that affect procrastinators' health.

If you are a major procrastinator, you likely have some cigarettes on you. At least, they are tobacco-based rather than cannabis (but you probably have had those too at some point). And what goes better with a cigarette than a drink, one that has a little alcoholic bite to it? Better not have too many—you don't want to pass out while smoking, because you haven't checked your smoke detector or changed its batteries in quite a while. And that wasn't a salad you had for dinner, not with all those calories. Well, if you got it at a drive-through, what do you expect? This brings up your driving. Have you noticed that most people are scared when you are behind the wheel? Don't get angry with me, though you do tend to get angry quite a bit. Don't you?[23]

In short, smoking, excessive alcohol use, drug abuse, recklessness, overeating, risky driving, and fighting, not to mention promiscuous sex, are all activities that procrastinators tend to do a little more of rather than a little less. They all tap into procrastinators' impulsiveness, making gratification the one thing they don't delay. If you currently partake in even half of these vices, you are not exactly a poster child for a healthy lifestyle. Odds are, your choices will catch up with you.

RELIGIOUS PROCRASTINATION

Despite being born in the fourth century, St. Augustine is interesting enough to this day that a musician, specifically Sting, has written a song about him. Prior to his conversion, St. Augustine was a follower of what was then the world's most popular religion, Manichaeism, and he knew the pleasures of the flesh way better than you would expect of any saint. Though Manichaeism was against procreational sex—which partly explains why it died out—it found recreational sex more forgivable, an option that St. Augustine and his multiple mistresses indulged in enthusiastically.[24] Their libidinous lifestyle more than explains how St. Augustine became the patron saint of beer or, at least, of brewers; it became his defining temptation. After converting to Christianity in A.D. 386, he had trouble turning his back on a woman's embrace, his most famous quote being, "Please lord, make me chaste, just not today!" He kept putting off celibacy, feeling utterly defeated by his procrastination.* Then one day, in his garden in Milan, he heard God directing him through a child's voice to "take up and read." He grabbed the Bible, which opened to this precise passage from St. Paul's Epistle to the Romans " . . . not in orgies and drunkenness, not in promiscuity and licentiousness, not in rivalry and jealousy. But put on the Lord Jesus Christ, and make no provision for the desires of the flesh." With such a direct message, he redoubled his efforts for a holier life.

St. Augustine's plight is a common one.† The world's great religions are tough on procrastination, universally viewing it as a detour from the path of salvation and enlightenment.[25] Their disapproval makes sense, because putting off good acts in order to sin will put you in spiritual jeopardy. Here are some samples that show how.

* In his own words: "And when Thou didst on all sides show me that what Thou said was true, I, though convinced of its truth, only repeated my dull and drowsy words, 'Right away, one minute, leave me but a little.' But 'right away' wasn't ever right now, and my 'little while' went on for a long while . . ."

† For another example, see St. Gabriel Possenti, who consistently swore whenever he became seriously ill that he would enter a religious order, only to have his resolve disappear when he healed. It took several bouts of illness before he kept his word, whereupon he contracted tuberculosis and died a few years later.

Hinduism, to start with, is defined by the *Mahabharata,* especially a section called the *Bhagavad Gita,* a religious text preached by the god-figure Krishna.[26] In it Krishna declares, "Undisciplined, vulgar, stubborn, wicked, malicious, lazy, depressed, and procrastinating; such an agent is called a Taamasika agent," unworthy of rebirth. In Islam, postponement of good deeds is primarily what the Arabic word for procrastination, *taswif,* refers to.[27] Similarly, *The Pillars of Islam,* the foundational book on Islamic law, has much to say on procrastination, none of it good.[28] The same is true of Buddhism, despite its often being taken to be the world's undemanding and unobtrusive feel-good religion. From the Pali Canon, the earliest written Buddhist scriptures dating from about the first century B.C. until today, the message has been consistent and clear.[29] As the American-born lama, Surya Das, says: "We have to stop procrastinating, pretending that we have forever to do what we want to do and be what we long to be."[30] But the religion in which procrastination appears to be the biggest problem, judging from the number of times it is mentioned, is Christianity. Sermons aplenty preach against procrastination, mainly because the faith emphasizes repentance.* People may lead a sinful, selfish life, but can seek forgiveness on their deathbed and still be redeemed, cramming for the finals so to speak.

Procrastination is a universal theme in all these religions because we cannot predict when we will die; thus, the best time to repent, to act morally, to commit ourselves to doing good is now.

---

* Such as Johnathan Edwards' eighteenth-century classic *Procrastination, or The Sin and Folly of Depending on Future Time.* Similarly, you have Reverend Edward Irving's, "Procrastination is the kidnapper of souls and the recruiting-officer of hell;" and Reverend Aughey's, "Procrastination has populated hell. All the doomed and damned from Christian lands are victims of this pernicious and destructive wile of the devil. It is foolish to procrastinate."

A parable from *The Mahabharata*, Hinduism's epic narrative, demonstrates this reasoning. The hero, Yudhisthira, promises to donate some money to a beggar *tomorrow*. His younger brother Bhima hears of this and runs out to ring the court's victory bells. "Why," asked Yudhisthira, "did you ring the bells?" Bhima replies, "To have made such as promise, you must have victory over life. Otherwise, who knows what tomorrow will bring?" Similarly, Sayyiduna Ali Murtadha, the fourth Caliph of Islam, wrote, "Everyone who is taken by death asks for more time, while everyone who still has time makes excuses for procrastination." If our clock suddenly stops, our souls may be damned if we put off good deeds, meditative practice, and requests for forgiveness.

The universal holy war, then, isn't against forces of darkness but against forces of nature, our own human nature. Religions are all battling procrastination among their believers and converts because whatever promised lands or promised rewards they offer will most likely be granted in the *distant* future. Inevitably, everlasting salvation is being deeply discounted against a backdrop of sins that provide pleasures immediately. The world may be spiritually divided by how we view God or the good, but when it comes to damnation, procrastination leaves no doubt that all religions have a lot in common.

THE PURSUIT OF HAPPINESS

If procrastinators tend to be less wealthy and healthy than the doers among us, it is likely that they are going to be less happy too. And they are. This is partly on account of the stress of procrastination, which frequently gives rise to guilt. It is not unusual for procrastinators to suffer more for putting off the work than they would have suffered by actually doing the work itself. Consequently, when they finally tackle the task, they are often relieved, admitting, "This isn't as bad as I thought." Rita

Emmett, in her *Procrastinator's Handbook,* considers this a law, which she codifies under her own name, as Emmett's Law: "The dread of doing a task uses up more time and energy than doing the task itself."

The online procrastination discussion boards often serve as confessionals of delay-induced torment. Here are half a dozen examples culled from two online forums, *Procrastinators Anonymous* and *Procrastination Support:*

- I've been very successful in many ways and managed to accomplish a lot in my life. But the process is miserable—I procrastinate, feel terribly guilty, get depressed, do work marathons, promise to change, and then start procrastinating again. I'm now at a point professionally where I've procrastinated so much on so many things that the work has really piled on and I'm fearful and unclear about how to dig myself out of the hole I'm in.
- The semester started two weeks ago and so far it has gone well. I was doing every assignment early and had so much free time but since then I have reverted to my old self. I fear for the worst and I have about two months until mid-term when my marks are due. I know I'm not as bad a student as shown in my report cards but I can't seem to get my work schedule in gear.
- Whenever I told people I was a horrible procrastinator, they would usually laugh, and then say they were too. But they seemed fine; their lives weren't on the brink of destruction because of their procrastination, like mine was. Can any of you please help me out?
- I really just want to DO WHAT I'M SUPPOSED TO WHEN I'M SUPPOSED TO DO IT! Whether I intrinsically want to or not, like NORMAL PEOPLE do. It hurts me so much that I cannot simply do that.
- And I'm so ashamed of even needing to resort to something like this. What kind of person am I that I have such a lack of self-control? I have fought and fought and fought over the years . . . I feel like it's a dying battle.
- This habit isn't funny, but I've always pretended it was. Really, though, it's pretty tragic. It takes me months to respond to e-mails, costing me personally, socially, and financially . . . the only thing I really ever finish is dessert.

Unfortunately for these procrastinators, guilt and poor performance won't be the entire story. When it comes to gratification, procrastinators stress *immediacy.* Like the spoiled rich girl Veruca Salt from *Charlie and the Chocolate Factory,* they don't care how, they want it now. Immediate gratification often comes at the cost of larger, later rewards, so consequently, procrastination is like running up a charge on your emotional credit card. You don't have to pay it now, but when the bill is finally due there will be compound interest. We fritter away the days with the small pleasures of television and computer games, of Internet surfing and Sudoku puzzles and end up with nothing to show for it. This is a recipe for regret.

In the short term, we regret what we do, but in the long term, we regret what we don't get done. Inaction causes us the greater suffering. Not to have done, not to have tried, to have put it off—this is part of the human condition, so we all suffer from it to some degree. You almost certainly have or will have regrets in at least one of these three life areas: Success, Self-Development, and Intimacy.[31] Looking back on our lives, it is common to feel that we should have gone for that degree or tried harder in class, that we should have mustered up the courage and risked rejection for that date, or made time for that phone call to Mom. We are haunted by the ghosts of our own lost possible selves—what we might have been: could've, should've, but didn't.[32]

I am no exception to procrastination's rule of regret. My brother Toby suffered from sarcoidosis, the same debilitating disorder that took the life of comedian Bernie Mac. When my family had to make the decision to take Toby off his ventilator and wait with him until he took his last breath, I was crushed with knowing what a fool I had been with my time. I regret putting off trips to see his plays. I regret not making it to the hospital sooner to see him. I regret the littlest things, like not taking more time to watch a bad movie on TV with him while eating take-out. He was the smartest, funniest person I have ever known and I took it all for granted. In keeping with life's synchronicity, soon after my brother's funeral I found a poem in the newspaper written by Mary Jean Iron. I clipped it to remind myself of my carelessness. It is still there, in my desk drawer, waiting for this moment:

Normal day, let me be aware of the treasure you are.
Let me learn from you, love you, bless you before you depart.
Let me not pass you by in quest of some rare and perfect tomorrow.
Let me hold you while I may, for it may not always be so.
One day I shall dig my nails into the earth, or bury my face in the pillow,
Or stretch myself taut or raise my hands to the sky
And want more than all the world, your return.

Put down this book and get going. Don't hesitate: call your mother, start writing that essay, ask out that special person you have had your eye on. Now is the moment you have been waiting for.

LOOKING FORWARD

Have you really put this book down? I didn't think so, but don't worry. I know it is not that simple. Interventions are still coming—you will hit them when you reach chapter seven. Right now, I want to continue focusing on the price of procrastination. In the next chapter, we look at the economic cost of procrastination to society. When we calculate the final figure, it will probably be larger than even your most outlandish guess.

CHAPTER SIX

The Economic Cost of Procrastination

HOW BUSINESSES AND NATIONS LOSE

Momentary passions and immediate interests have a more active and imperious control over human conduct than general or remote considerations of policy, utility or justice.

ALEXANDER HAMILTON

When exploring procrastination, no other country provides as many good examples as the United States. Almost two-thirds of all procrastination research is done with American citizens, and no wonder, given what it costs them. Here's how to calculate it. First, how many workers are there in a country? For the United States, the figure is over 130 million, but we will round down for ease of calculation. Second, what is the annual average wage for those workers? Estimates can reach over $50,000, but we will be conservative and go with the lower figure of $40,000. Finally, how many hours do people work each year? The Organisation for Economic Co-operation and Development provides that figure:

Americans clock in 1,703 hours, or slightly more than 212 eight-hour workdays, each year.[1] Finally, we have to determine how many hours each day people procrastinate. Two companies, America Online and Salary.Com, partnered together to survey the work habits of more than ten thousand people; the result was over two hours of procrastination in every eight-hour day, not including lunch and scheduled breaks. Once again, we will round the estimate for ease of calculation, this time downward to an even two hours.[2]

Keep in mind as we calculate the final figure that I've used conservative estimates at every step. We have 130 million people who spend about two hours out of every eight at work procrastinating, or 414 hours per year. Each hour is worth at least $23.49 (i.e., $40,000 divided by 1,703 hours), though if their employers are making a profit, they are worth more than that. At a minimum, then, procrastination is costing organizations about $9,724 per employee each year ($23.49 times 414).[3] Multiply that by the total number of employees in the United States, and you get $1,264,1200,000,000. In other words, a conservative estimate of the cost of procrastination for just one country in just one year is over a trillion dollars. This number may seem surprisingly large, but not if you are an economist. Gary Becker, who won the Nobel Prize for economics, concludes, "Indeed, in a modern economy, human capital [the work people do] is by far the most important form of capital in creating wealth and growth."[4] With a quarter of each person's work day spent dithering, procrastination is going to be costly.

Still, if this trillion-dollar figure makes you balk, fine. Revise any part of these calculations downward to what *you* think is reasonable. Cut the number of procrastination hours in half, pay everyone minimum wage, but pretty much anything times

130 million is still going to be a hefty sum. Myself, I think the true costs of procrastination are far more than a trillion dollars. Procrastination during the business day is only part of the picture.[5] Our ability to save money or make timely political decisions is also affected by procrastination, and the costs there should be over a trillion dollars too. And here is how it is happening.

TIME IS MONEY

The more we procrastinate at work, the more it costs us. Unfortunately, it's not just entry level workers who procrastinate but their managers and CEOs as well. Consider the Young Presidents' Organization, a club of corporate heads under forty-five who run companies worth more than ten million in revenue. In a survey of 950 of its members, the most troublesome problem reported was "facing up to a task which was, for various reasons, personally distasteful."[6] As my own research program shows, organizational teams, work groups, and task forces procrastinate.[7] The graph on the next page charts the average work pace of business groups over the course of their projects (the solid line) along with a hypothetical steady work pace (the dashed line). In both form and content, it parallels the graph from chapter 2 that featured student procrastination. As can be seen, both students and business groups demonstrate the same shape of curve, whereby people start off slow and then pick up the pace.*

---

* Group-level procrastination is actually common enough to receive an official name; many business academics call it *punctuated equilibrium.*

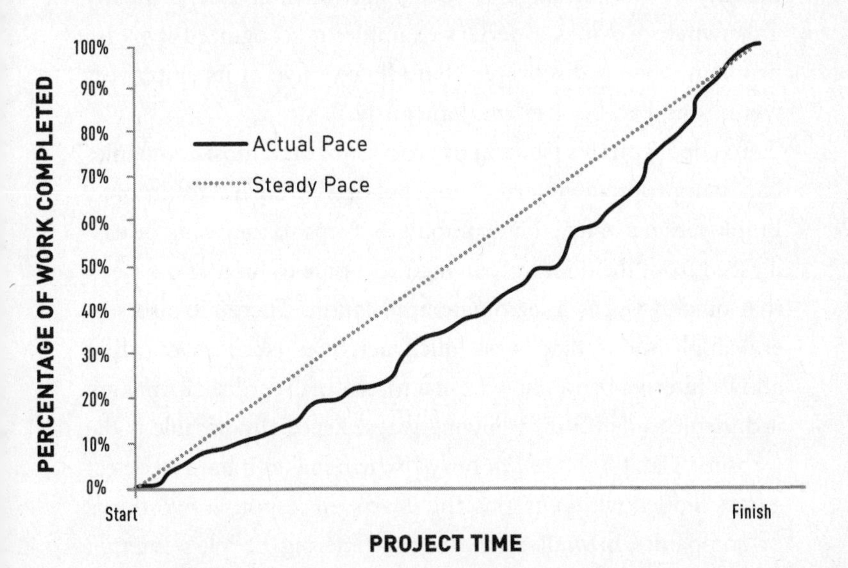

How has procrastination wormed its way into every inch of the business world? For the most part, by way of the same device that tempts students from their studies—the Internet. Dubbed e-breaking or cyberslacking, surfing the Net is the most serious of employees' time-wasting activities.[8] About one in four people admit to playing online games on the job. In fact, gaming websites report a sharp drop in traffic at exactly 5:00 p.m., the end of most people's work day.[9] Similarly, "video snacking," when people surf for and trade clips of all types, is a huge distraction. Though video use tends to spike during the lunch hour, it is prevalent at all times, and is expected to soon account for half of all Internet traffic.[10] As summarized by Miguel Monteverde, executive director of AOL Video, "Based on the traffic I'm seeing, our nation's productivity is in question."[11] Interestingly enough, this trend extends to pornographic sites as well, which get 70 percent of their traffic from the nine-to-five crowd.[12] Finally, of course, there is social networking. The company Talkswitch provides a perfect example; it recognized it had a problem when it discovered that all sixty-five of its employees were using Facebook—simultaneously.[13]

To cope with this tsunami of procrastination, most companies ban inappropriate Internet use, but it is difficult to enforce. Employees rearrange their computer screens so they can't be easily seen from the doorway, giving them time to hit a "Boss Key" that quickly opens a legitimate application. There are also several applications that mask illicit activities, such as one that allows Internet browsing within a Microsoft Word shell, making it difficult to detect dillydallying ways. Especially notable is the website "Can't You See I'm Busy," which makes it hard to detect games hidden within graphs and charts. In response, two-thirds of companies firewall their servers, fettering people's Internet access to various degrees. WebSense, which ironically makes software that filters the Internet, automatically monitors employees' Internet use and cuts off their access when they reach two hours of personal surfing. Other organizations enforce wide-ranging, perpetual restrictions on gambling, pornography, video sharing, and social networking sites alike.[14]

Banishing people from games and Internet sites does not eliminate polymorphic procrastination because it can manifest itself in so many ways. Solitaire is pre-loaded on most Windows platforms, making it the top computer game of all time, even favored by former president George W. Bush.[15] Memory keys often have games embedded on their chips, as do smartphones, which provide unrestricted Internet access. You can also go old school and avoid the computer completely. The ritual start of many a working day involves the diversion of the news. When I visit my sister, we scramble to be the first to get to the Sudoku in the morning paper. In the White House, Bill Clinton completed the *New York Times* crossword puzzle daily.

Procrastination isn't fuelled by games alone. As Robert Benchley quipped, "Anyone can do any amount of work providing it isn't the work he is supposed to be doing at that moment." We procrastinate on important tasks by doing the unimportant. For many of us, this means e-mail, which now takes up 40 percent of work life.[16] With every notifying "ding," workers instantly redirect their attention to reading the latest in an endless stream of electronic missives. Only a small seam of this e-mail bonanza is useful; the rest is junk. Though this deluge of electronic debris is partly composed of spam—unsolicited bulk e-mail—our greatest threat is the enemy behind the lines. Coined *friendly spam,* much of the junk we receive is created by our friends and co-workers who carelessly mass e-mail us about every social event, virus hoax, urban myth, trivia tidbit, or arcane corporate policy change. Since all these e-mails have the potential to be useful, they must be read to conclude they aren't. And then there are e-mail's peripheral effects. In a study of Microsoft workers, people took an average of fifteen minutes to re-focus on their core tasks after answering an e-mail interruption.[17] Combine this with the finding that information workers check their e-mail accounts over fifty times a day, over and above the seventy-seven times they text message, and theoretically no work should ever get done.[18] More realistically, the business research firm Basex puts the interruption and recovery time at a little over a quarter of the work day (about two hours),[19] which is consistent with studies on multi-tasking that conclude that switching attention is extremely detrimental to performance.[20] In short, despite the veneer of activity that e-mail checking provides, there is not much light for all that heat.

SAVING FOR LATER YEARS TOO LATE

Procrastination doesn't just diminish our wealth by decreasing our productive hours. It also reduces the benefit we gain from our productivity itself. Our wealth is determined not only by the money we make but also by the money we save. Saving is a tried-and-true path to riches, as every dollar you put aside starts to reap the miracle of compound interest. Furthermore, since the dollars you save are invested, savings can help the nation as a whole, spurring economic expansion. When adopted, a policy of savings can be hugely successful. Since 2004, average Singaporeans, for example, have been wealthier than the average American largely because they save more.[21] Unfortunately, when procrastination overtakes a society, saving becomes the exception and borrowing becomes the rule, a trend that can easily lead to financial ruin. Just consider your retirement savings account.

Aside from your plans to win the lottery, retirement rests on a three-legged stool. The first leg is the government, which, due to a bad habit of spending more than it receives, won't always be able to deliver on what little it promises. In the United States by the year 2040, for example, people can hope to receive only about two-thirds of their scheduled Social Security benefits, and on the heels of the 2008 global financial crisis, this percentage will probably decrease.[22] The second leg is represented by businesses, which can put money aside for you as a form of compensation, typically in the form of a Defined Contribution plan.* In such a plan, you decide how much, or rather how little, of your paycheck to contribute,

* These Defined Contribution plans are usually supported by the government and go by a variety of names, depending on the country you live in. For example, Americans have their 401(k)s, while Canadians have RRSPs. The UK has Pension Provisions, while France has Special Retirement Plans.

and most allocations are matched by the company. The third leg is you, your decision to initiate and open independent retirement accounts. This is the most dependable option—except, of course, that it still depends on you.

By becoming a society of procrastinators, we have caused the retirement stool to be increasingly wobbly, as most people are socking away less.[23] People are neither starting their own retirement accounts nor contributing to company plans, despite the fact that allocation matching is the equivalent of getting free money. When they leave work, their financial backsides rest on a stool supported by a single stubby peg derived from the government's forced savings program. Again, procrastination proves to be particularly poignant in the United States. In 2005, after decades of decline from originally double-digit rates, American household savings finally went into the negative. In other words, instead of saving today's money for the future, people were going into greater debt by spending tomorrow's money today—on average about half a percent more than they earned. To do this, not only did they borrow against their homes, in the form of mortgages, but about one in five borrowed against funds they had already set aside for retirement, putting themselves further behind.[24] Worst of all, some of this financing was arranged through "liar loans," which initially seem affordable but eventually create financial ruin. Variable mortgages entice homeowners to buy well beyond their means, while "pay day" advances provide the desperate with temporary respite but leave them much worse off. They end up repaying each loan many times over; the interest rates of "check cashing" shops often exceed 500 percent a year.[25] These are financial products that procrastinators are prone to fall for, products with short-term benefits but exceedingly high long-term costs.

The experts share the consensus that this situation is not ideal. At least something should be put aside for retirement; ideally, you should be saving 10 to 20 percent of your salary or higher if you are already in your forties.[26] Even before the 2008 global financial crisis, which alone lowered pension accounts by at least a fifth, an increasing number of Americans believed they were not putting enough aside for their old age.[27] And they are right. When retirement comes, more than four out of five Americans will find they haven't saved enough for their needs and by then it will be far too late to do anything about it.[28]

Retirement procrastination transforms the golden years into grim and gray poverty. It means living on skid row or with the kids, if you had them and if they'll have you. To prevent this from happening, governments have employed a few tricks. Tax breaks for contributing to registered saving plans are a good start, but to make the most of them, these breaks need to be accompanied by a definite deadline: that's what procrastinators respond to. Stipulating that retirement contributions must be put aside by tax time is an effective strategy, as it breaks down long-term retirement savings into a series of yearly goals.[29] Still, on its own, it hasn't proved to be sufficient, and so governments around the world are exploring another technique: automatic enrollment.[30] Applying the same negative-option marketing ploy used by mail order book clubs, employers can now *automatically enroll* their employees in pension programs with default investing options. Employees are free to withdraw or adjust their investment strategy at any time, but procrastinators will typically delay this decision, too. The result is a huge bump in enrollment.[31] Another neat trick comes from the trademarked *Save More Tomorrow* plan, developed by the behavioral economists

Richard Thaler and Schlomo Benartzi.[32] Rather than automatic enrollment, they use a strategy that exploits procrastinators' tendency to discount the future: employees can choose *now* to save *later.** That is, they must decide this year whether to start saving next year, and just as in automatic enrollment plans, once they have filled out the paperwork that commits them to saving, they will put off filing more paperwork to reverse their decision.

POLITICAL PROCRASTINATION

Governments, like people, have a bad habit of spending more than they receive. As I write this book, central government debts around the world are reaching commanding heights, often exceeding half the wealth their respective countries annually generate. By the time you are reading this book, it will be even worse. The United States, for example, will likely have finally hit the 100 percent mark, the point where it owes everything it makes in a year (that is, its total GDP). In dollar terms, that's an eye-popping $16 trillion. How did we get so deeply in debt? Governments display the same intention–action gap that defines all procrastinators: they form intentions to stop spending but change their minds when the moment to act is upon them. The United States has repeatedly tried to curb its own spending by legislating a borrowing limit—essentially reining in the government credit card.[33] Unfortunately, this is

* Furthermore, the amount put aside doesn't rely upon their present wages but draws solely on the extra money gained from assumed future salary raises. This is a nifty ploy, based on the "wage illusion." Wage increases typically keep salaries level with inflation, so you aren't really any richer. Still, a raise often feels like "extra money," instead of drawing on exactly what you are making now.

akin to an alcoholic locking the door to the liquor cabinet but leaving the key in the hole. Politicians simply vote away their previous debt resolution and install a new higher limit, a process they have repeated *hundreds* of times.

Governments are perpetually focused on quick fixes that solve the issues of the moment; the urgent displaces the important. This isn't a new insight. The American founding fathers understood this early on. I opened this chapter with a quotation from Alexander Hamilton, "Father of the Constitution," featured on every American ten-dollar bill. Similarly, James Madison, "Father of the Bill of Rights," wrote, "Procrastination in the beginning and precipitation toward the conclusion is the characteristic of such [legislative] bodies." And regarding the threat of debt specifically, here is a revealing quotation from George Washington: "Indeed, whatever is unfinished of our system of public credit, cannot be benefited by procrastination; and, as far as may be practicable, we ought to place that credit on grounds which cannot be disturbed, and to prevent that progressive accumulation of debt which must ultimately endanger all governments."

The American founding fathers were right; just take a look at the graph on the next page, which is similar to the two you have already seen, on student procrastination and the dillydallying of organizational teams. This one shows the average length of time it took the U.S. Congress to pass bills over the years from 1947 to 2000.[34] For every session in fifty, Congress passed the bulk of its bills toward the end of the session.

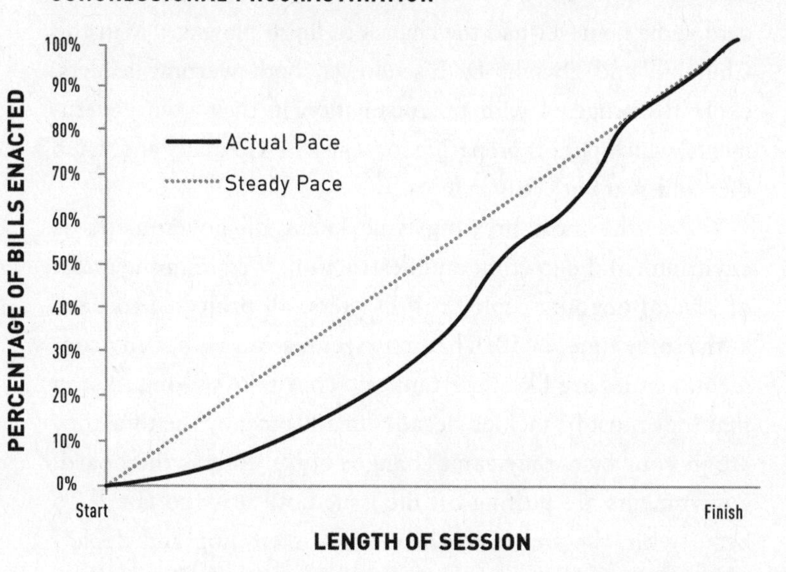

Though some bills are delayed due to political maneouvering, a large part of the delay should be due to procrastination. Furthermore, one can determine which groups are the worst procrastinators by comparing the surface area between the two lines—that is, between the steady work pace (the dotted line) and the actual work pace (the solid line). The more they are procrastinating, the greater the surface area. And Congress soundly beats out even the average college student when it comes to putting things off.

The result of all this procrastination is more than simply delay dealing with the national debt. All long-term national goals and challenges tend to be put off as well, no matter how threatening. The outcome of America's War of Independence was partly determined by procrastination. In a key battle, George Washington crossed the Delaware to destroy a Hessian garrison: Colonel Rahl, head of the garrison, actually had prior warning of the invasion but decided not to read the report until later, after a card game he never had the chance to finish playing.[35] Winston Churchill and Dwight D. Eisenhower, both wartime leaders, explicitly struggled with procrastination in their own governments, which put off preparing for war with Germany and, later, the Cold War with Russia.[36]

Today, the most pressing issue facing all governments is environmental depletion and destruction. We are in the midst of several ongoing ecological disasters, all projected to peak at the same time: 2050. That may seem far away, but environmental issues are like supertankers. They take so long to stop that they must be tackled decades in advance; by the time they are in your face, they can't change course. Across the board, governments are putting off the issue until it is too late.[37] To begin with, the soil beneath our feet is eroding and depleting.[38] With about 40 percent of agricultural land already damaged or infertile, what will happen in 2050 when the little remaining arable land must feed over nine billion people? It is also doubtful whether there will be enough fresh water to grow the necessary crops; the projection is that 75 percent of countries will be experiencing extreme water shortages by that same date.[39] The sea tells an almost identical story.[40] Approximately 40 percent of oceans are already fouled and overfished, with species disappearing around the world. But it won't get *really* bad until 2050, when the last of the wild fisheries are projected to collapse.

Interestingly—if that's the right word—these environmental disasters make the debate over global warming almost superfluous. With so many catastrophes projected, the consensus is grim. Even the futurist Freeman Dyson, who doubts global warming, concludes, "We live on a shrinking and vulnerable planet which our lack of foresight is rapidly turning into a slum." However, if climate projections hold true, we can expect about a three-degree increase in temperature by 2050.[41] No matter what country you are in, there won't be any place that will truly benefit from this change. Entire ecosystems, like the Amazon rainforest, are expected to collapse, about a third of all animals and plants will become extinct, and billions of famine refugees will fight to determine who starves to death first. Since many of us will be around in 2050, it is worth taking a private moment to envision what this tomorrow will mean to you.

Government bodies have been alerted to this possible future for a long time. In 1992, 1,700 of the world's leading scientists, including most Nobel Prize winners, signed the "World Scientists' Warning to Humanity," which stated in the most explicit terms: "A great change in our stewardship of the earth and the life on it is required, if vast human misery is to be avoided and our global home on this planet is not to be irretrievably mutilated." For even longer, we have known what to do about it. Unfortunately, we are procrastinating about translating this knowledge into action.[42] We could have avoided all these environmental issues if we had acted early. We can still mitigate them if we act now. The problem isn't informational or technological; it is motivational.

Still, be thankful that government procrastination isn't worse. Since the founding fathers of America were among the first to acknowledge the problem of procrastination, they did try to reduce its effects. Recognizing that what is expeditious can too easily prevail over what is wise, they tried to put temptation at a distance through *bicameralism:* legislation must pass through two houses or chambers. Using the exact terminology of "hot" and "cold" cognition favored by today's scientists, George Washington explained to Thomas Jefferson why they needed a senate as well as a house of representatives.

"Why do you pour coffee into your saucer?" Washington asked.

"To cool it," Jefferson replied.

"Even so," Washington said. "We pour legislation into the Senatorial saucer to cool it."[43]

Aside from Washington advocating a scandalous breach of etiquette (as "to pour tea or coffee into a saucer . . . are acts of awkwardness never seen in polite society"), it is a solid strategy that has been widely adopted.[44] What can be initiated immediately will hold much, much greater sway over tomorrow's better options. By purposefully building in delays, such as a senatorial house of sober second thought, the Constitution reduces the effects of time. Since it takes longer to pass all legislation, bicameralism focuses decision making on factors other than whether an aim is immediately obtainable. In other words, the added delay of a second house ensures that everything is going to take a while.

LOOKING FORWARD

We live in a world where our impulsive nature is only appreciated by those seeking to exploit it. But this is beginning to change. The field of behavioral economics, which recognizes our capacity for irrationality, is being incorporated into governmental public policy. Recently, the Gallup Organization hosted the inaugural Global Behavioral Economics Forum. Events like this have started to draw the attention of economic and political leaders from all shades of the political spectrum; both British Conservative leader David Cameron and U.S. President Barack Obama are exploring behavioral economic solutions.[45] Phrases from Obama's inaugural address highlighting this need for change appropriately resonate, especially our need "to confront problems, not to pass them on to future presidents and future generations." Some of this thinking has already been translated into action, such as legislation making it easier for businesses to automatically enroll workers in retirement savings plans. Still, much more needs to be done.

As individuals and as a society, we pay a hefty price for our procrastination and have done so since the beginning of history. But we can bring into millennia of dillydallying to an end today. A good start is to continue reading—the rest of the book is dedicated to actionable intelligence that puts putting off in its place. No matter what your procrastination profile—whether you lack confidence, hate your work, or are ruled by impulsiveness—there are proven steps you can take. And though we may have wished for this advice to have been available earlier in our lives, as we all know, working ahead of time is not really in our nature, is it? Perhaps we're now ready.

CHAPTER SEVEN

Optimizing Optimism

BALANCING UNDER- AND OVER-CONFIDENCE

$$\textbf{MOTIVATION} = \frac{\textbf{EXPECTANCY} \times \text{VALUE}}{\text{IMPULSIVENESS} \times \text{DELAY}}$$

A positive attitude may not solve all your problems, but it will annoy enough people to make it worth the effort.

HERM ALBRIGHT

I remember few darker days of the soul than those I spent hunting for a job during a harsh economy. Job hunting is humbling—and humiliating—and it tests you to the very core. As rejections and months of unemployment add up, a gnawing uncertainty makes you doubt who you are. When bills mount so does the pressure to settle for less, to take that job you swore was beneath you. But then, when you finally lower yourself to apply, you find that even that possibility is out of reach. Here is where the value of faith comes in, whether in yourself or in a God with a plan. Against all facts and experience, you have to believe that the next interview, the next lead, or the next day will bring a different answer. Belief in oneself separates the successful person from the procrastinator; without such confidence, the couch beckons, the television distracts, and dreams of the future become what could have been.[1] Many procrastinators doubt their ability to succeed and as a result, stop making the effort. Once effort disappears, failure is inevitable.

Beliefs are powerful because they form or directly affect *expectancy,* making them a motivational keystone of the Procrastination Equation. As you become less optimistic or less confident in your ability to achieve, your motivation also ebbs: the more uncertain you are of success, the harder it is to keep focused. This self-doubt is usually associated with novel and difficult tasks, but it can also become a chronic condition: expectation of failure. Poor self-perception then becomes a self-fulfilling prophecy—by expecting to fail, we make failure a certainty because we never dig in and make an intensive effort. Since beliefs can create reality, we need a healthy dose of optimism to motivate us toward success.

On the other hand, too much optimism can also lead to procrastination.[2] Remember Aesop's fable about the race between the Tortoise and the Hare? The far faster hare was so certain of his victory that he took a nap halfway through the race. The tortoise, moving slowly and steadily, overtook his slumbering competitor and won. As Michael Scheier and Charles Carver, psychologists who have spent their lives studying optimism, write: "It may be possible to be too optimistic, or to be optimistic in unproductive ways. For example, unbridled optimism may cause people to sit and wait for good things to happen, thereby decreasing the chance of success."[3]

Over-optimism is particularly prevalent when we estimate the time a task will take. It's called "the planning fallacy." Most people are not very good at predicting the length of time required for completing even commonplace tasks.[4] For estimating the time it will take to shop for Christmas presents, to make a phone call, to write an essay, the rule is "longer than you think." I myself am making edits to this very chapter far closer to my publisher's deadline than I'd like. We can't really help ourselves; it's a built-in bias of memory. To estimate how long future events take, we recall how long they took in the past. Our retrospection automatically abbreviates this time, and edits out much of the effort and obstacles. Unfortunately, this exacerbates the negative effects of procrastination. If you are leaving something to the last minute, there is actually far less time than that.

We need to find a balance between gloomy pessimism and Pollyanna optimism. Jeffrey Vancouver, a psychologist at Ohio University who specializes in the study of motivation, has succeeded in locating optimism's sweet spot. He found that, in a sense, we are motivational misers who constantly fine-tune our effort levels so that we strive just enough for success and use the prospect of failure as an indicator that we should up our game.* Look at the figure on the next page.[5] The vertical axis is motivation and the horizontal axis is optimism (that is, how difficult we perceive the task to be). Sensibly, we want the greatest reward for the least effort. Along the horizontal axis moving right, we start off with impossible tasks, too difficult to pursue. Why concentrate our resources where we will reap no reward? As tasks

* Sports teams constantly struggle against this trend, as it is natural to feel that last year's victory ensures the next season's success. As Bill Russell, winner of the NBA's most valuable player award five times over, notes, "It's much harder to keep a championship than to win one . . . there's a temptation to believe that the last championship will somehow win the next one automatically."

become easier and our optimism increases, we reach a tipping point. Motivation suddenly peaks: we believe that a win is possible, even though it will require considerable effort. As our optimism rises even further, our motivation falls, this time slowly. Eventually, we end up at the far right of the figure with tasks we believe we can easily perform. We're not motivated to accomplish these tasks because we deem them literally effortless. Most procrastinators are on the left of this chart, underestimating their ability, but a few are on the far right, believing that they are better than they really are.[6]

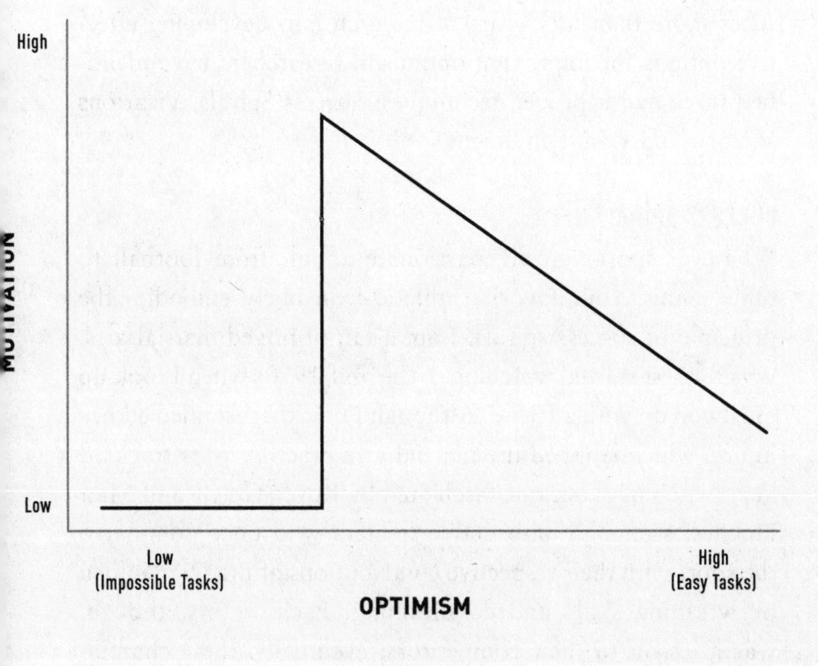

Since most procrastinators tend to be less confident than non-procrastinators, we will start off by focusing on how to increase optimism, as it plays a central role in expectancy. Then we will consider overly confident procrastinators and learn how to gently deflate their overblown expectations.

REALISTIC OPTIMISM

A little optimism helps us persist when it comes to tackling difficult tasks. "Next time," you might optimistically think, "it will happen for me." Such a belief will keep you going much longer than a more realistic "Success is going to take about two or three dozen more tries." But it isn't obvious how to achieve this sunny disposition. Slogans and aphorisms such as "Be positive!" tend to be as ineffective as they are popular; they work best for people who are already optimistic and can actually make matters worse for people who aren't.[7] But don't despair. After more than fifty years of research into developing effective options for improving optimism, researchers have identified three major proven techniques: Success Spirals, Vicarious Victory, and Wish Fulfillment.

SUCCESS SPIRALS

Whatever sport you are passionate about, from football to table tennis, your favorite athletic icon likely embodies the principle of success spirals. I am a fan of mixed martial arts, which I first started watching in the mid-1990s when I took up tae kwon do with a friend. Although I quickly sustained a knee injury, which stopped my martial arts practice in its tracks, I kept watching. I became fascinated by Royce Gracie and Matt Hughes, seemingly unbeatable fighters who once dominated the sport with their respective contributions of Brazilian jujitsu or wrestling skills and conditioning. Each victory, though, was a lesson to their competitors; eventually, these champions' abilities were countered or copied, and they fell. A titleholder of five years ago would likely be hard pressed to remain a contender today. One of the few champions who managed to endure is Georges St. Pierre. Remarkably, he attributes his present success to an old failure—he was knocked out by Matt

Sera. As St. Pierre puts it, "I think that loss was the best thing that ever happened to me, and skill-wise I'm way better than I used to be before." In a rematch between the two the following year, the referee stopped the fight when Sera was unable to defend himself from St. Pierre's attacks.

What makes Georges St. Pierre such a resilient combatant is his history of overcoming adversity, which includes a hardscrabble Montreal childhood. His persistence enabled him to transform initial failure into success, which in turn gave him the confidence to continue fighting and to improve in the future.[8] This is an example of a success spiral: if we set ourselves an ongoing series of challenging but ultimately achievable goals, we maximize our motivation and make the achievement meaningful, reflecting our capabilities. Each hard-won victory gives a new sense of self and a desire to strive for more. It is similar to the way Polynesian explorers colonized the South Pacific. From their home port they saw in the distance signs of a new island—a new goal—reachable if they made the proper provisions. Setting sail, they eventually made land, only to see another distant island from their new vantage point.[9] Every step forward is enabled by the step just taken.

For those who suffer from chronic discouragement and expect only failure, success spirals offer a way out. Initiating them is the trick, as everyday living doesn't easily provide a structured and confidence-building series of accomplishments. However, great opportunities are available: wilderness classes and adventure education. Much like tribe members in a season of *Survivor,* participants from management trainees to juvenile delinquents go on outings where they are challenged to overcome extremely difficult tasks with the help of inspirational guides. Outward Bound is the longest-running and most popular of these wilderness programs. In small groups, participants complete demanding expeditions on land or sea that can involve rafting, sailing, rock climbing, caving, orienteering, or horseback riding. Problem solving and personal responsibility are built in; individuals have to make key decisions, both before (what to pack?) and during (which way and how?). As a hundred studies have concluded, these wilderness programs improve self-concept, particularly self-confidence.[10]

One of the keys to the power of such programs is that participants leave with a vivid success experience they can hold on to—there's nothing vague about crossing a river or climbing a mountain or figuring out how to deal with the unexpected. Personal stories of triumph can bolster people's spirits for years to come. "I did it!" translates into "I can do it." In follow-up assessments, wilderness program participants report that their self-confidence kept growing; having accomplished in the wild tasks they thought they couldn't possibly do, they set higher goals for themselves at home. This is the essence of a success spiral: accomplishment creates confidence, which creates effort resulting in more accomplishment.

Parents can start these success spirals in their children. Structured extracurricular activities that provide a circle of encouragement and a venue for achievement can increase a child's academic achievement and self-esteem as well as reduce drug use, delinquency, and dropping out.[11] In particular, scouting provides an almost textbook recipe for creating tangible challenges that promote feelings of confidence.[12] With the motto "learning by doing," the Scouts reward a progressive series of tasks with proficiency badges that recognize each accomplishment, culminating in the coveted super-scout Baden-Powell Award.[13] Building a fire, setting up a tent, camping out, and cooking a meal for the group are all accomplishments kids can tell their parents about and—more importantly—remember themselves. Such success stories gradually build into a narrative that helps a child face the next challenge.*

Here is a personal example of a success spiral in action. A close friend of mine has a son with self-confidence and anxiety problems: since he doesn't expect to succeed, he gives up quickly. So, his parents enrolled him in martial arts at a very strict tae kwon do dojo. It took the boy several attempts to get his yellow belt, but eventually he did. This turned out to be the pivotal experience that changed the course of his life, and it wasn't because he became better at fighting. Every time he was tempted to give up in other areas of his life, especially school, his parents reminded him of how he had to persevere to get that yellow belt and how good it felt to receive it in the end. Having overcome obstacles in the past, he now routinely strives to overcome any new ones that arrive.

As adults, you might not have the time to try Outward Bound or share my passion for martial arts, and you are definitely too old for the Scouts. No worries; there are plenty of other options to create a success spiral. The secret is to start small and pay attention to incremental improvement, breaking down large and intimating tasks into manageable bits. Like the old adage about how to eat an elephant—one bite at a time—you carve difficult projects into a series of doable steps, purposefully planning in some early accomplishments. If you don't feel up to writing a whole report, find a small portion

* We could also cite the International Farm Youth Exchange or 4-H clubs (i.e., Head, Heart, Hands, and Health). With a similar slogan of "learn by doing," they also aid in youth development. Having branched out considerably from their agricultural beginnings, they actively prepare students to excel across a variety of specialties, especially the sciences. Ask any alumni of any 4-H club what they thought of it; overwhelmingly they will testify that it was a major contribution to their self-confidence.

you do feel capable of. Could you do the headings? Perhaps there are a few apt quotes to set aside? How about finding a few similar pieces to inspire you or to provide direction for organization? If you can't run a mile, then run a block. Stop when you've done that and next time try two blocks. Keep note of your progress, and watch how quickly you get to a mile. Nobody has to know about your small successes; keep them as your own happy secret and let them encourage you. The trick is taking the time to acknowledge incremental change, perhaps by recording your performance in a daily log.

Remember, there is always a path toward progress, no matter how small the increments. The better you are able to recognize subtle advances toward your goal, the more likely your confidence will continue to grow.[14] Success breeds success.

To help you put this into practice, throughout this chapter and the next two, I've included sections called *Action Points*. These sections give you pointers about how to put what you have read directly into action, easily and without delay. Here is the first.

1. Action Points for Success Spirals: Think of an area of your life of real interest and then strive to improve just a little beyond your present skill set. As your confidence builds, you can also try exploring life outside your comfort zone. Consider this list (and add to it):

- Volunteer for more responsibility, either at work or in your community. If it involves hard physical work, like building houses for the homeless, all the better. Those sore muscles will remind you of your effort and your success.
- Travel to a place you've always wanted to go but thought you never would. Give yourself bonus points if you don't speak the local language.

- Try an adventure course such as white-water rafting, mountain climbing, bungee jumping, or skydiving.
- Learn a new skill. Sign up for a class in cooking, kickboxing, photography, or music. As you advance, pay attention to the small improvements in your skills and recognize them as victories.
- Challenge yourself by pushing an old hobby to a new level. If you are a runner, train for a race; join an amateur sports league; or tackle the harder solos in *Guitar Hero*.
- Break down the tasks that daunt you into smaller and smaller pieces. Keep formal track of your progress. Count your successes.

VICARIOUS VICTORY

When I was a child, zoos were made up of cages, not habitats, and animals were truly captive. My father once took me to see the elephants. A mother elephant and child were on display side by side, both of their right hind legs secured to the ground. A large and heavy chain limited the baby, but the mother only had a slender rope. "Why Daddy?" I asked. "Shouldn't the big chain be around the big elephant?" No, he explained to me, the younger elephant needs the bigger chain because it is still struggling to become free. Eventually, it will accept that the chain won't break and, like the mother, it will stop trying. Once the baby elephant believes that it can't escape, the flimsy rope will be as effective as any cage.

Though I told it in the first person, this is a motivational story I've heard many times. Its implication is that we have untold strength, but that we were broken and tamed at some point and we don't realize how easily our potential could be regained if only we tried. I find it almost impossible not to be stirred by it, longing to break my own metaphorical ropes. There are many other motivational stories with this capacity to give us vicarious victory—from King Henry's "St. Crispin's Day" speech to Winston Churchill's "We Shall Fight on the Beaches." The most powerful of these are biographies of successful people that you can relate to.

Consider the effect one such story had on entrepreneur Kaaydah Schatten. Despite being raised in profound poverty by alcoholic parents, today she is a multi-millionaire and international franchise owner, a transformation she partly attributes to early inspiration. At a young age, Schatten read the life story of Catherine the Great and, seeing a common thread with her own heritage—Kaaydah is of a royal line, being the hereditary chieftain of the Quakiutl tribe—she adopted Catherine as a role model. To reap a similar benefit, perhaps you too can find the right story, another's life history that resonates with your own and speaks to your potential.

But people with extremely low self-confidence may need something stronger than inspirational stories to help them take their first step. Pessimists tend to put down any personal victories with a stream of negative self-talk: "Anyone could have done that," "It was all luck," or "It won't happen again."[15] They need active forms of encouragement to believe that their success is due to their own effort: that when they try, good things happen. We normally absorb encouragement of this kind through social support, peer groups, and role models. From adolescence on, our peer group is a determining factor of our own development.[16] Hang with the wrong crowd and they can hold us back. Hang with the right crowd and their successes can inspire us to think, "If they can do it, I can too!" Attitudes are catching, so you would be smart to hang out with groups of upbeat people. The social group we associate with helps cement our own view of what is possible and what we ourselves should strive to be. Giving up or continuing to strive—both are contagious.[17]

A few groups seem particularly well structured for fostering a positive spirit. Service clubs like the Elks, Masons, Rotarians, or Shriners have millions of members worldwide, all bent on doing good work for their communities, but your options don't end there. My wife goes to a local Calgary group, the Famous Five, which holds women's leadership luncheons. I'm indebted to Toastmasters, a club that promotes public speaking and is endlessly encouraging and welcoming. You can even start a group yourself.[18] Benjamin Franklin, for example, labeled his friends the Junto or the Leather Apron Club. Every Friday night, they would have a few beers at a pub and discuss how they could help their community.

2. Action Points for Vicarious Victory: Seek inspiration from stories or, better yet, from social groups. It is easier to believe in yourself if you are surrounded by others who believe in themselves—or you. Here are some suggestions:

- Watch inspirational movies. Here are a few I've seen: *Men of Honor, My Left Foot, Apollo 13, Invictus,* and *Hotel Rwanda.*
- Read inspirational biographies or autobiographies. The most effective will resonate with your own background, so use the bookstore staff to help you find an appropriate book. For example, if you are a chef, read *Humble Pie* by Gordon Ramsey, in which he speaks of his hard upbringing.
- Listen to inspirational speakers. Great athletes, heroes, and entrepreneurs regularly speak about their experiences. Seek them out.
- Join a community, service, or professional association. By hanging out with people who are trying to better themselves or the world around them, you will be infused with optimism.
- Start your own support group. As long as it contains a circle of mutually encouraging friends, it can be your running clique,

your religious study group, or in the case of Ben Franklin, your drinking buddies.

WISH FULFILLMENT

Professional athletes often use visualization to achieve their goals. Before going to sleep every night, they imagine the perfect golf swing or triple axel landing. The detailed mental recreation of a performance engages mirror neurons that engrave the act in your brain almost as deeply as if you were actually practicing it.[19] Visualization can also combat procrastination through the technique of *mental contrasting*.

The expert on mental contrasting is Gabriele Oettingen from New York University, who has made this technique a cornerstone of her psychology career.[20] Begin by imagining what you want to achieve. If it is a car, visualize yourself behind the wheel, cruising for all to see. If it is a job, see yourself in that dream career. Have you got a good mental picture? Good.

Now here's the all-important second step. Contrast where you want to be with where you are now. Visualize that dinged-up rust-bucket you drive or your dumbed-down joke of a job with its paltry paycheck. The result will be that your present situation becomes framed as an obstacle standing in the way of your dreams. Mental contrasting doesn't create optimism but it maximizes optimism's motivational benefits, creating energy and effort as well as jumpstarting planning. People who practice mental contrasting almost immediately start pursuing their dreams, putting a crimp in procrastination.

What happens if you forgo the second step and just focus on the positive fantasies alone? *Creative visualization* advocates just that. It involves creating vivid and compelling pictures of your heart's desire, with the aim of drawing this vision toward you. But Oettingen, who has researched this for twenty years, finds that such fantasies tend to have the *opposite* effect than advertised; they sap motivational energy.* The only wealth created by creative visualization is a rich fantasy life. Whether the task is preparing for exams, getting a job, recovering from surgery, smoking less, dating an attractive stranger, or improving personal relationships, she found that the worst-performing group used positive fantasies alone. You are better off not using the technique at all.[21]

3. Action Points for Wish Fulfillment: Fans of creative visualization don't have to stop what they are doing; they just need to add to it. Keep with the affirmations, the personal mission statements, but afterward reflect on where you really are. Here is a step-by-step walkthrough to make wish fulfillment work for you:

- Sit down in a quiet place and clear your mind. Think about the life you want for yourself.
- Break off a manageable piece of this future by focusing on just one aspect you desire. It may be a relationship, a job, a home, or a healthy body.
- Elaborate on all that makes this mental picture attractive to you. You can use a daily diary, create a collage of images, or just spend some quiet time concentrating on it.
- Then *mentally contrast* this future with where you are now. Focus on the gap. Put the same emphasis on vividly reflecting on this discrepancy as you did on imagining your idealized future.
- If, after mentally contrasting, you remain optimistic about realizing this ideal future, you will find more motivation to pursue

* Sigmund Freud much earlier drew a similar conclusion. Fantasy is primarily a process whereby we form an image of our desire and receive gratification from it alone. This is much like addiction to Internet pornography, where pixels take the place of people.

your goal. Procrastination will disappear as you start actively closing the gap between where you are now and where you want to be. You know what to do and have the drive to do it.

FANTASY LAND

Overconfidence is just as problematic as under-confidence. Forty-one days before the start of the Iraq war, the U.S. defense secretary, Donald Rumsfeld, estimated that it could "last six days, six weeks. I doubt six months." The allied troops would surely be greeted as liberators. The cost? It was supposed to be fifty to sixty billion dollars, not almost a trillion. Regrettably, military overconfidence leading to lengthy and unprofitable wars is quite common.[22]

In the business world, overconfidence creates a host of similar problems: mergers aren't usually finished within time or on budget.[23] Overconfidence, for example, contributed to the Concorde fiasco; despite mounting evidence that it wouldn't be profitable, Air France and British Airways continued to pursue its development.[24] Entrepreneurs often exemplify this point, reflecting Jeffrey Vancouver's observation that optimism has a sweet spot (see the graph on page 119).[25] Confidence is definitely needed to start a business, and entrepreneurs tend to have more of it than the rest of us. Just as the graph predicts, however, overly confident entrepreneurs tend to fail. When confidence becomes supersized and unearned, it fuels procrastination because the overconfident tend to discount serious problems and subsequently delay responding to them.[26]

Certain philosophies, such as the philosophy of Pangloss, a character created by Voltaire to epitomize naive and unrestrained optimism, exacerbate the problem of overconfidence. Over the last few centuries, unbounded positive belief has formed the basis of several success systems, such as Phineas

Quimby's *New Thought Movement* or Norman Vincent Peale's *Power of Positive Thinking.*[27] The best modern example of Panglossian thinking is *The Secret,* a book (and movie) developed by Australian television executive Rhonda Byrne. According to Byrne, thoughts have magnetic energy that draws like to like by a Law of Attraction—think positive and the positive will come toward you. There are millions of followers of this philosophy but I am not one of them.[28] The Law of Attraction separates positive belief from action, leaving belief free-floating and unconnected. It changes the story of the *Little Engine That Could* from "I think I can" to "I think it will." That's a big difference.*

To prevent ourselves from falling into over-optimism, we need a teaspoon of pessimism. As Freud put it, we need to activate the *reality principle:* to confront the reality of the situation when we are seeking the best way to achieve our goals. Invoking the reality principle is a sign that we have outgrown our childish and impulsive ways and can acknowledge the price we must realistically pay for our dreams. This entails imagining what could go wrong and how you would prevent or mitigate potential pitfalls. Neil Armstrong, the first man on the moon, used this principle during his lunar escapades. "Well," he would say, "I think we tried very hard not to be

* Little of this is new. Benjamin Franklin wrote about the need for hard work in *The Way to Wealth,* over 150 years before Wallace Wattles' *The Science of Getting Rich,* the book that inspired *The Secret.* Even if you adopt the premise that magical thinking works, it is traditionally thought to operate contrary to the way professed by *The Secret.* Magnets actually attract their counter, that is positive attracts negative. Consequently, boasting about or predicting a positive result means it is less likely to come true; we jinx the outcome by tempting fate. It is why we knock on or touch wood after reporting good luck or health, in an effort to avoid the curse and allow the good luck to continue.

overconfident, because when you get overconfident, that's when something snaps up and bites you."

In business, this reality check is a standard step of *crisis management*. Adages to this effect are well worn: "If you fail to plan, you plan to fail," or "An ounce of prevention is worth a pound of cure."[29] We can apply this principle to procrastination in two ways: Plan for the Worst, Hope for the Best, and Accept That You're Addicted to Delay.

PLAN FOR THE WORST, HOPE FOR THE BEST

Very few succeed in major life reforms on the first try; most of us need multiple attempts. Take New Year's resolutions, for example: it often takes five attempts before vows last for more than six months.[30] I myself sweated out several attempts to quit smoking before I successfully put cigarettes aside. For more serious alcohol or drug problems, the same need for repetition applies. Whatever you do, don't wallpaper over this painful and repetitive process; wishful thinking will only increase your procrastination.

Psychologists Janet Polivy and Peter Herman describe such dysfunctional over-optimism as the *False Hope Syndrome*. Overconfidence about the size, speed, and ease of major life changes is associated with lower success rates. If people have unrealistic, supersized expectations, they discount modest achievements. They *only* lost ten pounds. They smoked at a party. They skipped the gym for a week. They see these as "failures" and they lose momentum—they are more likely to give up and feel worse than before they made the resolution to change. This disillusionment is common, as the self-help industry instills incredibly high expectations and promises. If you are in the vast majority who don't transform as quickly as advertised, you feel that your failing is personal rather than a failure of the program.

Success requires balancing optimism with realism: it will be a hard slog and there will likely be lapses, but you can get back on track.[31] When I quit smoking, I paid attention to two variables: how many cigarettes I smoked during a binge and the length of time between binges. As long as the first number went down and the second went up, I was getting somewhere. Rather than believing you can entirely and easily beat the problem of procrastination, believe that you can beat it down. Instead of aiming to never procrastinate, aim to start just a little bit earlier on more and more projects. Modest gains can have significant outcomes. I have some students who start studying for exams only forty-eight hours beforehand, but if they started one day earlier, they'd increase their cramming time by 50 percent. As the author Louis L'Amour counsels: "Victory is won not in miles but in inches. Win a little now, hold your ground, and later, win a little more."

4. Action Points for Plan for the Worst, Hope for the Best: Life won't always go your way. Rather than expecting perfection, anticipate difficulties and setbacks. When they inevitably occur, you won't be as easily derailed. Here is how to inject some healthy pessimism into your plans.

- Determine what could go wrong to distract you on the way to your goal. Reflect honestly on your past experiences and seek counsel from others who have gone through similar difficulties. For example, take a look at the online forums about procrastination.
- Make a list of the ways you habitually procrastinate, and post it where you work.
- Avoid these pre-identified risky situations. For example, if text messaging is your problem, turn off your cell phone before you get to work.

- Develop a disaster recovery plan ahead of time. If you stumble and start skipping the gym, what emergency cord can you pull? Do you have a friend you can go to for a pep talk? Can you hire a personal trainer to get you back on track?
- If you find your motivation derailed, use your recovery plan. Focus on reducing the depth and duration of your motivational lapse.

ACCEPT THAT YOU'RE ADDICTED TO DELAY

When procrastination gets really serious, baby steps might not work. You may need a heavy-duty technique, one that you borrow from the Alcoholics Anonymous' twelve-step recovery program. AA's first step is: "We admit we are powerless over alcohol." Many find this admission a strange start to sobriety, since it runs counter to any notions of optimism and increases the chance that once you've had one drink, you will abandon all self-control and go on a bender.[32] Still, acknowledging powerlessness over alcoholism or procrastination can paradoxically lead to the elimination of both.

Indeed, it is possible to improve self-control by embracing your pessimism. How does this work? Well, truly acknowledging that any single failure of willpower inevitably leads to the collapse of all your self-control gives you far more motivation than believing that occasional lapses can be safely contained.[33] Abstinence is a preferable antidote to rationalizing every slip and indulgence. Since one drink, a lone candy bar, or a solitary cigarette is in itself inconsequential, we can trick ourselves into downplaying their significance. If we indulge in thinking one additional day of delay is always all right, then tomorrow's day of action never comes.[34] Maury Silver and John Sabini, who researched procrastination in the 1970s, describe this problem in terms of the prototypical student procrastinator:[35]

> Now suppose you had to decide what to do for just the next five minutes—either work on the paper or play one game of pinball. The paper can wait for one game—there is little long-term cost. In the short run, five minutes of pinball is far more pleasurable than five minutes of paper writing, and after all, how much of a paper can you do in five minutes? Pinball is the obvious choice. The game is over so you must decide about the next five minutes. The situation is only trivially changed, so you will reach the same result. Once you've taken the possibility of pinball seriously and fragmented your night into five-minute intervals, you may be doomed to play until you run out of money, the machine breaks, someone meaner than you wants to play. The trouble is, even five minutes has a real cost to the paper. Because a single game of pinball is brief, it is particularly seductive.

When it is time to decide whether to work or procrastinate, there is no shortage of excuses for giving in to temptation. Conditions will be better tomorrow, so I will start then; I'll work better after I get something to eat; it will be easier if I clean up first; I'll start after I finish this level, finish watching this show, finish this e-mail; this party/episode/diversion is especially good, so it would be unfair if I missed it; I deserve a break because I've been working so hard already; others are procrastinating, so why shouldn't I; it's just this once so it doesn't count; there's still plenty of time; and finally, it's already too late to make a difference so there's no need to start now. These are all justifications after the fact. Their only purpose is to assuage your anxiety and guilt.

There is only one surefire way to stop you from justifying your way into procrastination. Follow the Victorian era's greatest maxim: "Never suffer an exception to occur."* This is the same advice given by Alcoholic Anonymous. You buttress your commitment to early starts by believing that any slip will be catastrophic, that the initial step toward procrastinating is merely the first link in an endless chain. The specifics of tomorrow will be much the same as today: you will be tempted to incur a small but cumulative cost to gain a moderate immediate pleasure. If you decide to delay even once, your decision will be replicated daily and the consequences will grow. It is as these verses from Goethe's masterpiece, *Faust,* indicate. Asking for more time can be a deal with the devil:

> *Lose this day loitering—'twill be the same story*
> *To-morrow—and the next more dilatory;*
> *Each indecision brings its own delays,*
> *And days are lost lamenting o'er lost days,*
> *Are you in earnest? seize this very minute—*
> *Boldness has genius, power and magic in it.*
> *Only engage, and then the mind grows heated—*
> *Begin it, and then the work will be completed!*

5. Action Points for Accepting That You're Addicted to Delay: If you find yourself chronically procrastinating, consistently able to fool yourself into extended delays by finding moment-by-moment excuses, this may be the technique you have been

* This is from William James' 1890 textbook, *Principles of Psychology.* James is actually summarizing a recommendation made four years earlier by Alexander Bain: "It is necessary, above all things, never to lose a battle. Every gain on the wrong side undoes the effect of many conquests on the right." For that matter, what James considers the second greatest Victorian maxim is also relevant: "Seize the very first possible opportunity to act on every resolution you make.

looking for. Procrastination has a very deep hold on you, and to defeat it you need to accept this humbling fact.

- Take a moment to reflect upon how many times you have talked yourself out of your plans and into trouble. Start keeping a daily log to track your procrastination habits.
- Acknowledge that your biggest worry is your own weak will, that you *will* try to fool yourself into thinking "just this once."
- Accept that the first delay allows you to justify all the subsequent ones. By doing so, you will be far less likely to take that first step.

LOOKING FORWARD

This chapter is primarily for the low-expectancy Eddies, those who need just a little more confidence to reach their potential. Back in chapter 2, Eddie stopped believing in his ability to make the sale and because of this self-doubt, his failure became inevitable. If he had attended more to his progress, he could have initiated a success spiral. If he had supplemented this momentum by attending a sales support group, creating a little vicarious victory, he could have maintained a career in selling. You might have also stopped believing in your ability to advance your career, your personal life, or your health. You make plans to change but no longer truly believe in your ability to pull them off. Take a moment to review your results from the self-assessment quiz in chapter 2. If you scored 24 or below on the Expectancy scale, maybe you, like Eddie, should take a closer look at the techniques presented here.

On the other hand, there are a few of you who are overconfident, and you may be equally at risk. Confidence or optimism turns out to be a lot like vitamin A: too little of it will lead to blindness but too much of it can kill you. The trick is to find the sweet spot between being pessimistic and being happy-go-lucky, a place where you have faith in your ability to succeed but not so much faith that you fail to put in the effort. Whether you need your positive expectations fired up or dampened down, you are in luck. All the techniques shown here are rock-solid and scientifically sound. They will start working immediately and you will get better with practice. Believe me.

CHAPTER EIGHT

Love It or Leave It

FINDING RELEVANCE IN WORK

$$\text{MOTIVATION} = \frac{\text{EXPECTANCY} \times \textbf{VALUE}}{\text{IMPULSIVENESS} \times \text{DELAY}}$$

If time flies when you're having fun, it hits the afterburners when you don't think you're having enough.

JEF MALLETT

To warm up my students for a class on motivation, we play a game called *My Job Is Worse Than Your Job.* Since misery loves company, it is a lot of fun. We try to find the worst past employment experience in our group and then deconstruct the job to determine why it was so terrible. The room echoes with groans of sympathy as the students talk about summers spent shoveling pig manure or their scorching, exhausting, and mosquito-plagued months spent tree planting. But invariably, the jobs voted "worst" aren't the physically demanding ones; the worst are the mind-numbingly boring ones. For example, one bright young man used to spend his potential straightening cardboard boxes when they occasionally became misaligned crossing conveyor belts. I was once a lifeguard at a waterslide park, assigned to watch the same few meters of splash zone for time without end.

These sorts of jobs turn us into clock-watchers who wait for each agonizing minute to pass.[1] Since every aspect of the job has been mapped out, we are left with little to say about when or how to do the work, little chance for initiative or innovation. We must repeat the same actions endlessly. Are we doing a good job? Nobody really knows except when we slip up. Movies such as *Modern Times* and *Office Space*, in which the protagonists escape such employment purgatory, become cult classics. More recently, the award-winning television show *The Office* has been a success in half a dozen versions around the world. Part of the show's charm is its ability to demonstrate how humanity manages to rise above the squelch of meaningless work. Factory and office jobs, however, weren't always like this.

We owe the "modern" workplace largely to Frederick Winslow Taylor, the originator of scientific management.[2] Before he came along, the majority of work was skilled and somewhat immune to direct management, performed by craftsmen who learned their trade through years of apprenticeship and specialization. Managers couldn't easily supervise such artisans when they had little idea of how they did their jobs and it wasn't in the workers' interest to tell them.[3] Taylor's breakthrough was to fragment work into more easily managed elements—simple routine tasks lacking autonomy. When his system, Taylorism, was first implemented, back in the late nineteenth and early twentieth century, it was considered an abomination that lobotomized the human spirit, robbing work of its meaning and pleasure. People hated it so much that its introduction at the U.S. government arsenal at Watertown, Massachusetts, spurred a strike that led to a special investigation by the U.S. House of Representatives. The congressional committee concluded that man should naturally resent "the introduction of any system which deals with him the same way as a beast of burden or an inanimate machine," and took action to prevent the adoption of Taylorism in governmental facilities. When the industrialist Henry Ford implemented a similar system, employee turnover at his automobile factories increased almost tenfold; workers stayed barely a month before leaving. Taylorism, however, had an ace in the hole: it was efficient and it was profitable. Though Ford eventually had to double wages to fill his factory jobs, the improvement in efficiency allowed him to simultaneously increase his workers' pay and cut the cost of his Model T cars by almost half. In the end, Taylorism helped to give rise to cheap goods and a wealthy middle class that could purchase them. Assembly line work, on the other hand, still often sucks.

The tasks we hate are among those we tend to postpone. Because Taylor's system leads to standardized, repetitive, and rigidly controlled tasks, hating work can be a chronic state of being, an inevitable result of jobs being designed around mechanistic instead of motivational models.[4] What can we do about this? We might dream of returning to a time when what we desired to do and what we needed to get done were the same, but that's not realistic. Even if you are your own boss and can dictate your own terms, you still need to accomplish some tasks that are no fun, and these are exactly the jobs that people put off. Perhaps it's time to think about tricking yourself into getting them done. As the title of this chapter goes, you love it or leave it—until later that is.

GAMES AND GOALS

Whoever we are, we are likely to put off doing whatever we find excruciatingly dull. Boredom signals that what we are doing is irrelevant, and so the mind slides off the task.[5] It makes sense, then, that procrastinators are much more likely than non-procrastinators to perceive life's daily tasks as drudgery. Of all the boring tasks that fill the world, the one that tops most people's hate list is routine paperwork. The busywork—filling in timesheets, submitting expense reports, and supplying the data that companies and governments endlessly require—seems pointless, even when it isn't. Remember Michael Mocniak, that general counsel who got fired for putting off completing $1.4 million worth of invoices? Fortunately, however, boredom isn't inherently part of any job—anything can be made more interesting simply by the way we treat it.[6] Tom Sawyer, for example, managed to get the village boys to *pay him* for the privilege of whitewashing his Aunt Polly's picket fence. How? By insisting that they couldn't help and making them envy an unenviable chore. Here are a few effective techniques for turning leaden tasks into golden ones.

To relieve task boredom, try making things more difficult for yourself. (Don't overdo it though—when tasks are too difficult, frustration can take hold.)[7] Finding the balance between the difficulty of your task and your ability to do it is a key component for creating *flow,* a state of total engagement.[8] Flow states don't happen naturally, since many jobs are structured around an unvarying level of difficulty, whereas most workers' ability increases with practice. When work is new and its difficulty exceeds your ability, anxiety rises as you fumble to perform. Then, as you improve, the work can become engaging, but this motivational fit is momentary. When you have gained true mastery, boredom becomes the rule; you have done it all before. To prevent this descent into dullness, game playing becomes a common strategy. Set your own standards, create your own feedback, and try to beat your score. Can you do it in half the time? How about one-handed? Eyes closed? The comedy group Broken Lizard devised a film, *Super Troopers,* around this theme: five Vermont state troopers find ways of weaving games and shenanigans into their work, making their days passable.* One elderly potato-chip factory worker kept her days full by collecting unusual chip deviations that resembled famous people.[9] Competitive swimmers keep boredom at bay by imagining sharks in the pool water.

By the way, I can't help but notice that you are continuing to read this book despite the shelves of other books to choose from. I'd guess that procrastination is a problem you or someone in your family experiences, and that as a consequence you are finding these pages personally pertinent and interesting. You could put the book aside, but relevance keeps you reading. This is equally true for other actions and tasks: the risk of procrastination diminishes when tasks are relevant, instrumentally connected to topics and goals of personal significance.[10] Actions that don't fit self-determined and self-defined goals are *amotivational.*[11] They are imposed upon us and we reluctantly comply. At my university, we have many managers voluntarily coming to school each evening after working a full day at the office in order to earn their MBAs. I imagine their motivational chain of objectives goes something like this:

- They read the book so they can prepare for a test.
- They prepare for the test so they can ace the course.
- They ace the course so they can get the grades.

---

* For those who have seen the movie, "Meow."

- They get the grades so they can receive an MBA.
- They get the MBA so they can get a promotion.
- They get the promotion so they can make more money and enjoy their work.

All the sub goals in this hierarchy are predicated on the last—getting a promotion so as to enjoy more interesting work.[12] You need a string of future goals that you find intrinsically motivating to hook your present responsibilities onto. Break this motivational chain at any point and you leave it anchorless; goal commitment is negligible and, like a balloon, attention floats away with every waft.

The relevance factor is a major reason why procrastination decreases with age. As we mature, we increasingly connect the dots, seeing reasons for what we once thought was pointless. If you lack large resonating goals—life tasks—then your purpose now is to find them. It is a big world and you need to experience at least some of it. In the meantime, I will give you a generic goal that will inject any task with more meaning. Frame what you are putting off as a test of your willpower, and, as a buttressing side bet, tell your friends of your intention to start early. The goal of staying true to yourself and portraying your consistency to others will increase the pleasure of sticking with the task and resisting tempting alternatives.[13] For example, Barack Obama's public announcement that he intended to quit smoking helped him put cigarettes aside, with only the occasional lapse.[14]

To further maximize your intrinsic motivation, frame your long-term goals in terms of the success you want to achieve—an *approach* goal—rather than the failure you want to prevent—an *avoidance* goal. People who generate positive long-term goals go on to procrastinate less and perform better.[15] Advice like "Don't fall!" to the precariously balanced or "Don't forget the lyrics!" to singers increases the likelihood of the very outcomes they profess to prevent. Consequently, "I really want this book to get good reviews" is better than "I hope not to be openly mocked for my writing." Thinking "I want her to like me" is better than "I don't want to be rejected again." Almost any goal can be flipped from avoidance to approach, from what you don't want to happen to what you desire.[16] Just look at the following table:

	AVOIDANCE GOALS ARE . . .	APPROACH GOALS ARE . . .
1.	*Not* staying home	Exploring the World
2.	*Not* being tired	Having energy
3.	*Not* staying in a dead-end job	Finding your calling
4.	*Not* struggling with bills	Making more money
5.	*Not* leaving the glass empty	Filling the glass up
6.	*Not* beginning late	Starting early

On which side of that table do you usually reside? Do you focus on not eating treats when dieting (an avoidance goal) or on eating healthy meals (an approach goal)? Do you think about not procrastinating (an avoidance goal) or about starting earlier (an approach goal)? I thought so. So the lesson is: Stop making avoidance goals!*

1. Action Points for Games and Goals: It is said, at least by people quoting Shakespeare, that there is nothing good or bad in this world but thinking makes it so. The Bard is exaggerating a bit but he is essentially right. Frame your tasks appropriately; the way you view them significantly determines their value.

---

* Oops! Make approach goals!

- Avoid boredom by making tasks more challenging. Games can be handy here, with the rules limited only by your imagination and common sense. For example, when you are competing against your colleagues, almost any task can become a race to finish first or to get the most work done. In competing against yourself, you could also try to finish the task in fewer hours.
- Connect tasks to your long-term goals, to what you find intrinsically motivating. For example, if you are a social person, you could frame cleaning your house as "Providing an inviting home for family and friends."
- Frame your goals in terms of what you want to achieve rather than what you are trying to avoid. For example, think "I want to succeed" instead of "I don't want to fail."

ENERGY CRISIS

When I moved to Minnesota to work on my PhD, my wife, Julie, and I managed to snag a dream apartment: a converted warehouse loft. Rent was low—a key feature on a student income—and the location was close to both my university and her workplace. Even better, only a wide golden field separated us from the Mississippi River. Nothing, however, is all good. That field was full of ragweed, which triggered my hay fever. My allergies had never been bad enough to warrant medication, but after my third box of tissues, I quickly opted for an over-the-counter allergy drug. Suddenly, I couldn't get out of bed in the morning without repeated sharp prods from my wife. Work became incredibly difficult, like running through deep powder snow. What was wrong with me? Was I depressed? Overwhelmed? Finally, I read the back of the medication box: "May cause drowsiness." Later I learned that most allergy medications contain antihistamines, which have the same active ingredient as the sleep-aid *Nytol*. I was taking the equivalent of sleeping pills and no wonder I was having trouble tackling tasks.

Whether tiredness is drug-induced or not, being too tired is the number one reason given for procrastinating; 28 percent of people claim, "Didn't have enough energy to begin the task" as the cause.[17] When you are tired at the end of day, after your job has already got the best part of you, cleaning out the garage is the last thing you are going to do. Fatigue increases task-aversion, saps interest, and makes the difficult excruciating.[18] Whether it is an exhausted muscle or an exhausted mind, you can feel the burn of being burnt out.[19] When you're tired, it becomes even harder to force yourself to tackle jobs you dislike. Burnout saps your willpower because the exercise of will—self-control and self-motivation—takes energy. Whenever you have to suppress a competing impulse, you exhaust your energy stores and willpower. If you have to stop yourself from eating that cookie, you deplete your willpower. If you suppress an emotion, like laughter or anger, you deplete your willpower. If you are coping with stress, your willpower depletes. This decrease in self-control occurs after you make difficult choices, one reason why clothes shopping can be an exhausting ordeal if you lack an innate fashion sense. Those bizarre outfits that languish in your closet were likely purchased toward the end of a shopping trip.

To some extent, we should accept that we don't have infinite mental energy and acknowledge our motivational limitations along with our physical ones. Everyone understands why you can't run back-to-back marathons but it's not so obvious that equivalent internal struggles can be just as onerous. Perhaps we have trouble with procrastination because we demand too much of ourselves in a day, and it's possible that pursuing a less stressful, slower paced life would help us get energized.

Regrettably, we don't always have a choice. So what can we do when our "get-up-and-go" has "got-up-and-gone"?

Recognizing that our energy reserves are limited, we can strategically refuel and allocate them. You don't want to ever completely exhaust yourself; when you are sapped, you are likely to give in to your impulses. That is why dieters shouldn't let themselves get hungry, because they are likely to satiate themselves with the simple carbohydrate and fat combinations that saturate our world. Ironically, sweet treats will restore willpower just long enough for you to regret the indulgence.[20] So shield yourself from distractions by using moments of strength to enact other longer-lasting self-control techniques, especially distancing yourself from temptations.[21] This is the beauty of offices. Once purged of temptations, an office can become a temple of productivity, a place where following up on your intentions to work takes a lot less willpower.

Facing the challenges of writing a report at the end of the day, when you're already wiped, isn't the best idea either. You want to tackle it when you have the most zip, and when that occurs depends upon your circadian rhythm.[22] Some of us are morning larks, relentlessly chipper and active early in the morning, filling gyms in the pre-dawn hours. Others are night owls, slow starters whose energy levels peak later in the day. Night owls are more likely to be procrastinators, with a chronobiology best suited for after-hours endeavors; forcing themselves into an unnatural schedule, they gulp down caffeine in the morning in order to wake up, and alcohol in the evening to wind down.[23]

Whatever your rhythm, schedule that report writing to start a few hours after you wake up; it's when your mind operates at maximum efficiency, a period that lasts about four hours.[24] If you woke at seven in the morning, for instance, your peak performance likely occurs between ten and two, not really that wide a window. But if you clear your desk, turn off your e-mail, and shut your door for those hours, you can get an amazing amount of work done. You can extend this efficiency phase with a brief nap, twenty minutes or so, but if you're in an office environment, that's usually not possible. Still, a quick walk around the block can also refresh you around lunchtime. In any event, it's smart to start shifting toward less creative, more routine work in the late afternoon: you are losing IQ points by the hour. When you finally get home, the only decision you might be able to effectively make is whether to wind down with a glass of wine or a pint of beer. The good news is that the timing is perfect; twelve hours after waking is when your liver best metabolizes alcohol.

Finally, a typical pattern that many of us fall into when stressed is to cut back on exercise and sleep and make up for them with diet and stimulants, usually sugar, caffeine, and nicotine. In the short-term, this can be an effective energy strategy, but in the long-term, it will leave you worse off. Not only do stimulants lose their effectiveness with repeated use, but they can make exercise and sleeping even more difficult to achieve. As quality of concentration is gradually swapped for quantity of effort, you work longer hours while producing less, eventually working late into the night when you should be sleeping. These are bad energy habits.

You probably already know what you should be doing to solve these problems. Committing to a regular schedule of exercise has been shown to decrease procrastination.[25] Since many people in North America aren't getting a good night's sleep, I also recommend you start learning about sleep hygiene, which prevents people from polluting their bedrooms with the stress of the day, maintaining it instead as a sanctuary for escape.[26]

Sleep hygiene is the only thing that worked for my wife, who comes from a family of chronic insomniacs.

2. *Action Points for Energy Crisis:* Being too tired is the top reason for procrastination. Your energy stores are both a limited and a renewable resource, so actively replenish them and allocate your efforts wisely.

- Reserve your morning and mid-day peak performance hours for your most difficult tasks.
- Don't let yourself get hungry. Graze on small nutritious snacks as needed.
- Make time for exercise several days each week.
- Make sleep predictable, going to bed at the same time each night with a regular wind-down routine.
- Respect your own limitations. If after all this, you still are too tired to tackle your responsibilities, try to cut back on your commitments or get help completing them.

YOU SHOULD SEE THE TASK I'M AVOIDING

The sun sets and long shadows disappear into the darkness. Eyes dilate to adjust, but still the blackness obscures: uncertainty shrouds us and anything could emerge. Vulnerable now to the limitless unknown, we feel a suffocating fear. With night comes the time of monsters. Pull the blankets over your head and don't say a word: this is about survival . . . at least it used to be. Like three-quarters of kids, I grew up afraid of the dark, a dread largely passed on from my ancestors.[27] When nighttime was truly dangerous, that fear of ghouls and ghosts kept children quiet, stationary, and safe. Imaginary fears were an adaptive part of any culture.[28] The Northern Inuit teach their children of the Qallupilluit, which kidnaps children who walk too close to cracks in the ice, while the

Japanese have the Kappa, water creatures that eat urchins.* Maybe we can conjure our own monster to scare off procrastination as well.

The technique of productive procrastination might employ such a monster. It is a well-established ploy, advocated by no less than Sir Francis Bacon, the seventeenth-century philosopher and statesman. He proposed that we try to "set affection against affection, and to master one by another; even as we use to hunt beast with beast." We see productive procrastination in action when people spend precious hours sharpening pencils, scrubbing stoves, or cleaning bedrooms as an imminent deadline towers over them. Though by all outward appearances they seem suddenly afflicted by obsessive-compulsive disorder, such procrastination isn't entirely a waste of their time.[29] Things are getting done—though not quite the right things.[30] Psychoanalysts would consider it an example of displacement, whereby we shift impulses into a related but less threatening outlet, like picking a fight with a friend after being upbraided by our boss. Behavioral psychologists would point out that we are willing to pursue any vile task as long as it allows us to avoid something worse.

Productive procrastination isn't perfect—it reduces the cost of dillydallying but doesn't eliminate it. Rather than doing nothing useful while avoiding the big project, you are at least taking care of minutiae, "robbing Peter to pay Paul." It isn't as constructive as tackling the real work, but it does clear your plate and puts you in a much better position to dig in when you're ready. Sooner or later, though, you will have to face that monster you have been avoiding.

---

* Which they start feasting on first through the anus. It makes for a nice bedtime story.

3. Action Points for You Should See the Task I'm Avoiding: Don't let the perfect—never procrastinating—get in the way of the good—productively procrastinating. Meet your procrastination impulse halfway. By engaging in productive procrastination, you put off one task only to spur yourself toward tackling another.

- Identify a target task that you ideally should be doing now but have been putting off.
- Identify tangent tasks that also should be done and are *relatively* more enjoyable than your target task. You might be putting these off too.
- Accept the trade-off of avoiding the target task by tackling the tangent tasks. When you eventually get to the target task, you will be in a better position to complete it.

DOUBLE OR NOTHING

We are all too familiar with guilty pleasures. You know, the ones you indulge in after a long day doing things for others, after the kids are fed and in bed, the dishes done, and you finally get an hour to yourself. You slip off your work clothes, step into a robe, pour yourself a drink, and watch . . . oh yes, reality TV. Ah, the sweet cerebral abyss of spoon-fed entertainment. We all have the ability to self-reward, whether it be with a trashy book, a bowl of ice cream, or a luxury purchase. So let's put this talent to good use.

A principal problem with procrastinators is that they tend not to reward themselves after completing a task, often failing to appreciate their own hard work.[31] They give themselves no whispered kind word or planned treat after a task well done. Too bad, as such rewards are the easiest to implement and personalize. The specifics of soothing self-talk or a deserved indulgence will differ from person to person, but the effects remain the same. Whether your catchphrase is a silent "Atta boy!" or a "You go girl!" a little internal self-praise is a costless incentive for overcoming a challenging task. Similarly, whether it is a fine meal or a full vacation, a self-administered reward can pull us through the drudgery of work toward a project's completion. Even better, they offer motivational dividends, realized during subsequent endeavors.

This technique is called learned industriousness: people can learn to love their work.[32] You see, the enjoyable emotions generated by self-praise and other rewards tend to creep backward into the effort itself. That is, activities take on the attributes of their goals and can become rewarding in themselves. Money is the principal example of this phenomenon, having been instilled with value by virtue of what it can later buy. Hard work, by virtue of the achievement it can later generate, can be similarly infused, making such effort rewarding in the moment. Consequently, successful people find themselves in a virtuous circle: the anticipated rewards from winning help make the work more enjoyable, and that enjoyment helps them to win. With the future flavoring the present, they savor victory long before it is realized. It is a very nice arrangement, but the trick is in how to get it started. It may take a number of effort–reward cycles before the effort itself takes on the taste of the later reward.

While waiting for learned industriousness to kick in, you can enhance the pleasure of work in a more direct way: blend bitter medicine with sweet honey.[33] Try to find a compatible pairing between a long-term interest and a short-term impulse. If you combine an unpleasant task with one you find more enjoyable, the mixture may be enough to get you going. Getting together with a workout partner can spur you to exercise. Treating yourself to a specialty coffee can help you focus on your time sheets or your budget. But this method has its risks as well. Engaging a partner to help you finish a report or prep for an exam, for example, can degenerate into an evening-long bull-session with little learning to show for it. Still, the principle is sound. In the Adam Sandler movie *Billy Madison,* the title character has to redo his entire schooling, twelve grades in twenty-four weeks, to receive a sizable inheritance. In desperation, he engages an attractive tutor, who for every correct answer he gives, removes an article of her clothing.

4. Action Points for Double or Nothing: Take the time to recognize and reward your progress. Though success itself will eventually make effort enjoyable, right now you can artificially graft a little pleasure onto most tasks.

- Make a list of rewards you can administer to yourself, such as self-praise, frivolous purchases, or a night out.
- Promise yourself these rewards upon completion of the task you have been avoiding.
- Consider ways of making tasks more enjoyable, such as listening to music, sipping a specialty coffee, or working with a friend.
- Make sure that what makes the work more enjoyable, like partnering, doesn't override the work itself.

LET YOUR PASSION BE YOUR VOCATION

Perfect work exists, tasks people would do even in the absence of a paycheck. One example is gold farming.[34] Gold farmers are professional video game players who have become experts in massive multiplayer online role-playing games (MMORPG) like *World of Warcraft, RuneScape,* or *Star Wars Galaxies.* With their honed skills and long hours of play—at times, eighteen hours a day—they gain virtual gold and rare items that they then sell to other players for real cash. As documented by Ge Jin, a University of California PhD student and independent filmmaker, these professional gamers blur the line between work and play in a constructive way. Jin admits he was "shocked by the positive spirit there, the farmers are passionate about what they do, and there is indeed camaraderie between them."[35] Most telling is what many gold farmers do in their spare time—they continue to play.

Apart from the problem of who would buy all this make-believe money, gold farming isn't and can't be for everyone. Still, it captures the Holy Grail of job design, marrying high performance with job satisfaction. And it illustrates that finding work you want to do is a major step toward avoiding procrastination. Being intrinsically motivated by your job means you are rewarded simply by doing it; no need to delay gratification here. This combination can make work almost addictive; motivation shoots upward stratospherically, souping up creativity, learning, and persistence.[36] Speaking for myself, I love learning about motivation and I willingly work hard at it. Finding work you love is tricky, but let's try.

Finding your perfect job is at least as difficult as finding your soul mate. With almost 50 percent of marriages in our culture ending in divorce, the challenge is a tough one. With love, we seek the person who complements us; with work, we seek the job that could become a calling. In either case, a satisfying match is known as congruence, and it can be darn difficult to accomplish. The best predictor of love is familiarity brought about by physical proximity,* a good recipe insofar as it keeps

* As Sir Peter Ustinov concluded, "Contrary to popular belief, I do not believe that friends are necessarily the people you like best; they are merely the people who got there first."

down travel costs while dating.[37] At work too, we gravitate toward the best of our known options, not the best of all possible jobs. Expanding our world and improving our career choices is not a simple matter. We need to better understand ourselves and what different jobs can offer us, and then find a way to link the two.[38]

For most people, finding themselves and their calling is an ongoing struggle. If we all went with our first impulse, the working world would be primarily composed of firefighters and ballerinas. If we followed the dreams of our teenage years, we would mostly be professional athletes, fashion designers, or rap stars. Ask college students, and many of them want a career in film. On the other hand, choosing sensibly to be a doctor or a lawyer doesn't always pan out either; these were the initial career paths of Graham Chapman and John Cleese before they created Monty Python's Flying Circus. Most of us have to search for a calling while we are already working, deepening the commitment to a current and perhaps inappropriate career path. We may need the help of a matchmaker or, as we call them in the world of work, a counseling or vocational psychologist. These professionals assess your personality as it relates to work, typically relying on an assessment tool that divides interests into six themes: realistic (doing), investigative (thinking), artistic (creating), social (helping), enterprising (persuading), and conventional (organizing).[39] Jobs are profiled too, with firefighting falling under "realistic" and ballerina under "artistic." Vocational counselors will point you toward a variety of job choices, though it is up to you to at least try "dating" them. The assessment on the next page is my own, completed when I was seventeen years old. One profession clearly dominates my profile, one that requires a combination of investigative and artistic interests—a professor. I did not take these findings seriously at the time, but drifted about for a decade before finally coming to the same conclusion. Blame my strong need for autonomy.

INVESTIGATIVE			30	40	50	60	70
	THEME	MODERATELY HIGH	58				
BASIC INTERESTS	SCIENCE	AVERAGE	52				
BASIC INTERESTS	MATHEMATICS	AVERAGE	49				
BASIC INTERESTS	MEDICAL SCIENCE	AVERAGE	54				
BASIC INTERESTS	MEDICAL SERVICE	AVERAGE	50				

CODE	OCCUPATIONAL SCALES	Standard Score M	Standard Score F	Very Dissim.	Dissim. 12	Mod. Dissim. 21	Mid-Range 27	Mod. Sim. 39	Similar 45	Very Similar 54
IR	Veterinarian	R1	43							
IR	Chemist	25	31							
IR	Physicist	36	26							
IR	Geologist	39	39							
IR	Medical Technologist	21	28							
IR	Dental Hygienist		32							
IR	Dentist	44	37							
IR	Optometrist	47	34							
IR	Physical Therapist	33	44							
IR	Physican	40	46							
IRS	Registered Nurse	43	S1							
IRS	Math-Science Teacher	28								
IRC	Math-Science Teacher		24							
IRC	Systems Analyst	15	35							
IRC	Computer Programmer	33	32							
IRE	Chiropractor	34	37							
IE	Pharmacist	28								
I	Pharmacist		38							
I	Biologist	39	35							
I	Geographer	31	43							
I	Mathematician	32	25							
IA	College Professor	52	49							
IA	Sociologist	40	45							
IAS	Psychologist	34	36							

As in seeking love, there is more involved in finding your calling than identifying what you desire. Though a certain job could be the one for you, your feelings may not be reciprocated. Some jobs are out of our league because they are already being pursued by an excessive number of applicants. Supply and demand is harsh, and there may not be a demand for what you supply. Fortunately, there are plenty of other jobs that you could like just as much. The O*NET program in the United States catalogues nearly a thousand jobs, identifying those that are in demand by employers and that fit your profile.*

After accounting for your personality and for the job market, you will also need to consider your abilities.[40] Can you do what the job requires? Firefighters and ballerinas need to be athletically gifted, ranking in the top positions of the physical fitness category. If you want to be a rocket scientist or a brain surgeon, you'd better be blazingly smart. Linking individual abilities, such as stamina or mental capability, with the world of work isn't easy.[41] For example, I can tell you that if you are five feet tall, you shouldn't foster aspirations for a career in the NBA. But most of the time it isn't obvious whether you are following a dream or pursuing a lost cause. Just be aware that you want to find work that you not only love but have the capacity to excel at too.

5. Action Points for Let Your Passion Be Your Vocation: Not everyone has job mobility. Some are tied down by obligations and economic constraints and have to make choices based on security or availability. If you have the gift of choice, don't blow it! For the

---

* Go to: http://online.onetcenter.org/find/descriptor/browse/Interests/. If you check it out, look up my profession, Industrial-Organizational Psychologist. You will see that in addition to researching motivation, we also counsel workers about their careers.

next little while, finding a compatible fit between what you do and who you are should be an ongoing occupation.

- Look at careers involving activities you love or like doing.
- Filter out all the occupations for which you don't possess (or aren't willing to learn) the necessary skills or abilities.
- Rank the remaining careers by what is in demand. The harder the economic times, the fewer your choices will be.
- If you need help answering any of these questions, find a reputable career service for employment advice.*
- Start job hunting!

LOOKING FORWARD

In chapter 2, Valerie Without Value hated to write and put off her municipal politics assignment for so long that what she produced was second-rate. Instead of working, she indulged in the far more pleasurable acts of texting her friends and binging on video snacks. Hers is a regrettably common story, especially among writers.† To stop procrastinating, Valerie needs to find a way to heighten the value of her work. Connecting it to her greater career goals would be a good start. By identifying the type of writing she wants to do and framing the present task as a stepping-stone toward this goal, she should enact strategy elements of both *Let Your Passion Be Your Vocation* and *Games and Goals*. Also, she could have started earlier in the day, when she had the most energy, instead of toward the end, when her willpower was weakest (see *Energy Crisis*). And at the very least, she could have tried *Double or Nothing* and used that

---

* One suggestion is Career Vision, which focuses on both job success and satisfaction: http://www.careervision.org/

† For example, Douglas Adams, the bestselling author of *The Hitchhiker's Guide to the Galaxy*, had a legendary ability to avoid writing. As he quipped: "I love deadlines. I like the whooshing sound they make as they fly by."

municipal politics piece to motivate her to get other work done, procrastinating productively rather than cyberslacking.

If you scored 24 or above on Valerie's scale about value from chapter 2, you probably can relate to her life, though your problem might lie elsewhere than in writing.[42] If so, reviewing the techniques in this chapter would be a good idea, as there is indeed some wiggle room in the world to find work that suits us better and to fashion this work into something we love (or at least like). Let's transmute those motivationally inert and tiresome tasks into golden goals that engage you. Just think, it might even be fun!

CHAPTER NINE

In Good Time

MANAGING SHORT-TERM IMPULSES AND LONG-TERM GOALS

$$\textbf{MOTIVATION} = \frac{\textbf{EXPECTANCY} \times \textbf{VALUE}}{\textbf{IMPULSIVENESS} \times \textbf{DELAY}}$$

He that has not a mastery over his inclinations, he that knows not how to resist the importunity of present pleasure or pain, for the sake of what reason tells him is fit to be done, wants the true principle of virtue and industry, and is in danger never to be good for anything.

JOHN LOCKE

Impulsiveness is the last cause of procrastination we will address, despite its overwhelming desire to be first in all things. "Now, now, I want it now" is its mantra. If we have an inner child, this is it, and it wants that candy right away. Impulsiveness runs through every vice that involves weakness of the will. Not only does impulsiveness form the core of procrastination but it is strongly connected to dysfunctional relationships, lousy leadership, suicide, substance abuse, and violence. In their groundbreaking book *A General Theory of Crime,* criminologists Michael Gottfredson and Travis Hirschi argue that most misdeeds and misdemeanors are due to impulsiveness alone.[1] What inevitably happens when vices give more immediate satisfaction than virtues? The most impulsive person will be the most corruptible.

Consequently, impulsiveness stands at procrastination's centerfield, and has a much more intense relationship with procrastination than with any other personality trait. Whereas low self-confidence (expectancy) and propensity for boredom (value) have definite roles in creating procrastination, they are not in the same league as impulsiveness. Impulsiveness multiplies the effect of delay, making it a major determinant of the Procrastination Equation's outcome. A person with twice the average level of impulsiveness as a typical person will generally let the deadline become twice as close before starting to work. Unfortunately, if you are impulsive, you will always be somewhat susceptible to putting life off. Though you will experience a modest decrease in impulsiveness as you age[2] and not all situations will trigger impulsive action,[3] you can't escape your fate. Impulsiveness is not something you have, but something you are.

So what can we do about a chronic lack of self-control? Civilization has been chewing over this problem for thousands of years, figuring out how to tone down the limbic system and pump up the prefrontal cortex.[4] Since every generation has to rediscover these solutions in their own words, it is time for us to revisit and reframe a little ancient wisdom. Let's go back to the beginning of the Greek empire, its legendary poet Homer, and his epic *The Odyssey.*

COMMIT NOW TO BONDAGE, SATIATION, AND POISON

Known as Odysseus or Ulysses, this King of Ithaca reigned more than three thousand years ago, but is widely remembered to this day. In the battle to retrieve the beautiful Helen, it was Ulysses who thought up the famous Trojan horse, a giant wooden statue in which forty Greeks were hidden. Since the phrase "Beware of Greeks bearing gifts" was still hours away from being coined, Troy accepted the peace offering, only to have Ulysses and his men descend from the horse's belly behind their lines. For us, the most important of Ulysses' stories happens afterward on his sea voyage return. In a poorly planned itinerary, he fights dozens of monsters—the Cyclops, giants, drugged-out hippies known as lotus-eaters—but most important of all, the Sirens. These beautiful women, despite being perpetually naked and available, are unattached for good reason. They sing and their voices are so pure and captivating that they are irresistible; enthralled by their melody, you will want nothing but to listen and will blissfully starve, die, and rot. What do you do? Fortunately, on one of his previous stops, Ulysses had met the goddess Circe, who gave him some handy advice: fill his men's ears with wax to make them deaf and bind himself to the ship's mast so he could hear the irresistible song but couldn't act upon his urges. The bondage worked and Ulysses traveled on.[5]

How does this apply to us? Consider Ulysses' situation in terms of the Procrastination Equation in the chart on the following page. On the vertical axis, we have Ulysses' desire, showing that he always acts on what he wants most. On the horizontal axis, there is the time dimension, starting off on the left with the way he feels right now and then moving to the right, tracking the way his desires change over time, especially as he approaches the Sirens and then Ithaca. Initially, he wants to go home to

Ithaca, surprise his wife, Penelope, after his twenty-year absence, and slaughter all the suitors vying for her hand—as represented by the dashed line. He is noticeably less enthusiastic about dying at the hands of the Sirens, as represented by the solid line. However, his preference reverses when he reaches the island of the Sirens, where briefly the solid line peaks above the dashed line. If he hadn't taken Circe's advice and protected himself and his crew, they would have all stayed and died on the island. This is exactly what the Procrastination Equation predicts. As you get closer to a temptation, your desire for it peaks, allowing the temptation to trump later but better options. This probably happens to you all the time.

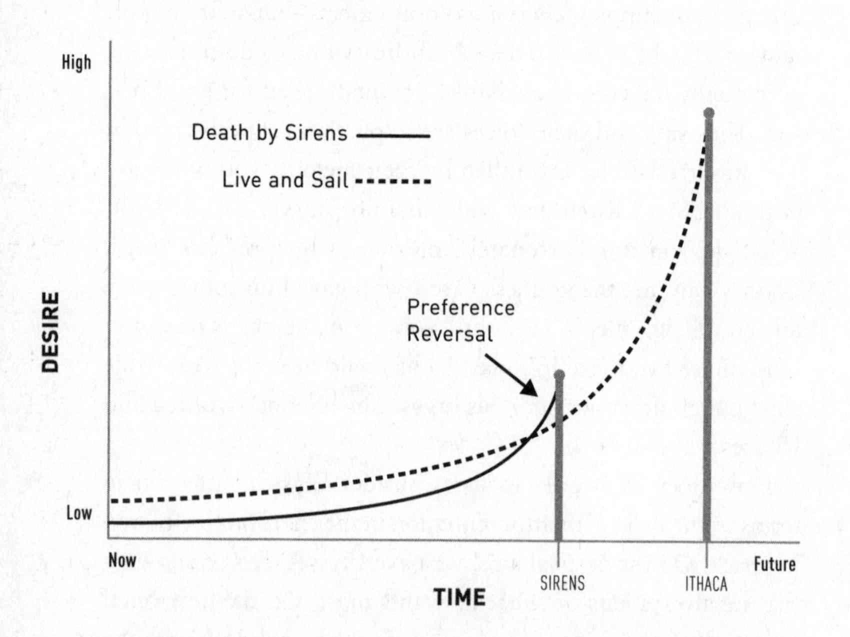

Right now, I'm sure you have no shortage of long-term goals: you want to lose ten pounds, stop smoking, get out more, or work harder. Maybe you want to start saving money for retirement or just for a trip. Standing between us and our aspirations are our Sirens. Instead of beautiful bare-breasted babes, they are the dessert cart, the television, or the amazing videogame. We wake in the morning with a clear desire to hit the gym in the afternoon only to succumb to the succubi of the immediately available. We want to diet but when some apple-crumble cake wafts under our nose, our willpower crumbles too. But if you can anticipate these powerful temptations, you can act in advance to ward them off. You can use the concept of *precommitment*.[6]

Because he heeded the warning about the Sirens, Ulysses acted before the urge was upon him, precommitting now to prevent himself from later weakness. Because he followed Circe's advice, Ulysses lived to sail another day. Unfortunately, we don't have our own goddess to warn us of our Sirens; it is notoriously difficult to anticipate our own temptations in the moment. Using economic terminology, *sophisticates* acknowledge their self-control problems, while *naïfs* are caught unaware by sudden shifts in their inclinations.[7] Most of us are *naïfs*, unable to fully anticipate how we will feel when cravings leap upon us.[8] In biological terms, our prefrontal cortex and limbic system just don't get each other, so we tend to underestimate the power of our own arousal—the heat of the moment—whether it is hunger, anger, or sexual excitement. And we forget the degree of regret we will feel after acting on these urges. Looking groggily into the mirror the morning after, we are mystified by exactly what our limbic system was thinking the night before.

Though we might be slow learners in regard to the power of our temptations, we do eventually learn. Give it some thought. When you are procrastinating, what are you doing? Do a few specific distractions come to mind? Can you name your Sirens? If so, let's start precommitting. Keeping true to your goals can be a limited time offer, so here's how to act now.

THROW AWAY THE KEY

A common military strategy to prevent your ships from being captured is to destroy them yourself, but such destruction has another purpose. The Spanish conquistador Hernán Cortés scuttled his own ships after landing in Mexico by filling them with water, even though the enemy was not yet in sight.[9] Similarly, William the Conqueror burned a few of his boats symbolically and had the rest dismantled when he made landfall in England.[10] In both cases, these men profited from their decision and went on to establish new dominions. Cortés destroyed the Aztec Empire and took their ruler Montezuma hostage. William's conquest of England ensured that the native-born nobility were replaced by those of Norman origin for centuries to come. By eliminating the means of retreat, they left their troops no option but to win, a strategy that dates back thousands of years. Sun Tzu summarizes it in his sixth-century text, *The Art of War:* "Throw the troops into a position from which there is no escape, and even faced with death they will not flee. For if prepared to die, what can they not achieve? Then officers and men together put forth their utmost effort."

Applying this principle to procrastination, we can also shield our long-term goals from immediate temptations. Our ships in this case are our alternatives, which we try to eliminate. Herman Melville reportedly had his wife chain him to his desk while he wrote *Moby-Dick.* To keep writing, Victor Hugo had his servant strip him naked in his study and not return with his clothes until the appointed hour.[11] Knowing that I will devour half the Halloween candy ahead of time, I don't buy it until hours before the trick-or-treating starts and take leftovers to the office for my colleagues the day after. Smokers, attempting to quit, give their packs away, telling friends not to lend them cigarettes. Revelers going out to the bar leave their credit cards at home and bring limited cash so they don't break their budgets.*

Unfortunately, as with so many of the strategies we have already encountered, precommitment can be difficult to enact, especially on your own. Ulysses had his crew to tie him up to the mast, but we usually find ourselves without sailors at our command. Technology is beginning to fill this gap. A few years ago, I was interviewed for an article in *Newsday* celebrating the fiftieth anniversary of the snooze button.[12] The snooze button is the devil's device, a procrastination-enabling technology that lets you easily put off your original goal of waking up, in order to grab a few more minutes of low-quality slumber. To counteract this temptation, people hide their alarm clocks across the bedroom, or make use of Clocky, a clock on wheels that, after you hit the snooze button once, bolts off your nightstand and beeps and flashes like a robot in distress. A number of similar applications have been developed for the computer. Google has the "Take a break" button, which disables your e-mail for fifteen minutes. Another feature is Mail Goggles, which prevents late-night drunken e-mailing by requiring you first to solve simple math problems after 10:00 p.m.[13] Others are being constantly developed, including a wide selection of add-ons for the Internet browser Firefox (*MeeTimer, LeechBlock*); for Apple users, there is the *Freedom* program, which will block your access to the Internet for up to eight hours. Unfortunately, most elaborate commercial time-control software, such as *Chronager,* is based on the idea of parental control instead of self-control; once you have the system of checks and balances

* Another great example is from Tony Wilson in the movie *24 Hour Party People.* Tony was a Manchester music mogul and aficionado of punk rock. Despite his success, he never retained much money. His explanation is pure precommitment: "I have protected myself from ever having to sell out by having nothing to sell out."

set up, you will need a friend to surreptitiously change your password and keep the new one a secret.

Despite their usefulness, however, such precommitments aren't entirely effective. Most of these examples merely make succumbing to temptation difficult but still not impossible. The crux of the problem is that the same cunning you employed to set them up is now turned against you; indeed, you are your own worst enemy. You can always run to the store to buy another treat, reformat your computer to get around nanny software, and throw pillows to suffocate Clocky. Samuel Coleridge hired thugs to prevent him from frequenting opium dens, only to fire them when the urge came upon him once again. In *Trainspotting,* Ewan McGregor's character nailed himself into a room so he could quit his heroin habit, only to extract himself later with the same determination.[14] More realistically, the mechanism at work here is delaying—not preventing—your access to temptations. As the delay lengthens, with luck the desire for the temptation is reduced in strength. A bowl of ice cream might beckon if placed within arm's reach, but its voice is muffled when shut inside the freezer. Naturally, the greater the desire for the vice, the greater the distance required to silence it.

SATIATION

Have you ever gone to the grocery store hungry? Bad idea. You likely wheeled down the aisles, filling your cart with indulgences that weren't on your list. Unpacking the bags at home, you loaded your cupboard and freezer with goodies that took you weeks to plow through and added pounds to your midriff. Really, all you needed was a small treat but in your state of deprivation you impulsively bought yourself a sizable feast. The pearl of wisdom, aside from "Never grocery shop on an empty stomach," is that more basic concerns must be attended to before concentration can be applied elsewhere.[15] Abraham Maslow, the father of humanistic psychology, based his theory of self-actualization on this insight, positing that we have a hierarchy of needs whereby basic, more visceral desires, like food and safety, must be attended to first.[16]

To precommit using satiation, we try to meet our needs in a safe and managed manner before they intensify and take control. If your appetite becomes too extreme, you will gorge yourself in seeking to satisfy it. Two common precommitment strategies are having a glass of water and garden salad at the beginning of a meal and grazing on small healthy snacks throughout the day.* A rather fun way of encouraging fidelity is to make love before your partner leaves for a prolonged trip, endorsed by no less than St. Paul the Apostle.† Smokers use the nicotine patch to reduce their cravings while heroin users take methadone. A broader use of this strategy is to schedule your recreational activities in your calendar first. Then pencil in your chores. Called an "unschedule," it can breathe energy back into life's grind.[17] In all these cases, the idea is to let off a little steam before our boilers burst.

TRY POISON

Even though registration deadlines are posted months ahead and reinforced with early bird discounts, the crush of applications

---

* Gastric surgery or stomach stapling is a more drastic form of satiation precommitment in that it reduces the amount of food needed to feel sufficiently suffonsified. That there is a non-negligible chance of dying during the procedure emphasizes the desperate measures people are willing to undertake to combat their desires.

† "Do not withhold yourselves from each other unless you agree to do so just for a set time, in order to devote yourselves to prayer. Then you should come together again so that Satan does not tempt you through your lack of self-control." (1 Corinthians 7:5)

for anything from training courses to 10K races typically occurs just before the deadline.[18] No surprises here. Presenting at a conference in New York a few years ago, I met Victor Vroom, an expert in leadership and motivation. Crossing Times Square with him, I noticed that neither of us had managed to secure rooms in the main hotel because we had both registered too late. Procrastinators, however, are paradoxically not always the last to sign up; sometimes they are the first. In an effort to precommit, they sign long-term health club contracts, buy season subscriptions to the symphony, or request home delivery of highbrow films from DVD movie clubs far in advance.[19] By acting now, they hope to irrevocably force their future selves to do what their present selves are unwilling to pursue, even if it means poisoning other alternatives.

A one-time common form of this precommitment device was the Christmas Club.[20] Invented by the Carlisle Trust Company in 1909, banks offered low-interest savings accounts that penalized you for early withdrawal. Despite today's easier access to credit, variations on Christmas Clubs still exist.[21] Why would anyone use them? Because they want to be under the threat of punishment: without the looming penalty, they fear they will withdraw and spend their money prematurely, and have nothing but good intentions to leave under the tree. The same principle can be useful in preventing weight gain. Weight Watchers is an international company designed to punish people for putting on the pounds. It provides assistance and advice for getting to and maintaining a target weight. Once you are firmly established at your ideal size, you receive a free Lifetime membership. But there is a catch. You must weigh in once a month and if you are more than two pounds over, the membership fees are reinstated until you again shed the pounds. I've also heard of a Danish chain of gyms that offers membership free of charge as long as you show up once a week.[22] Fail to exercise regularly and you have to pay.[23]

With the help of a merciless friend or perhaps an agreeable enemy, you too can raise the stakes on any venture. Just make a painfully large bet that you will lose only if you put off striving toward the goal you want to attain. Economists John Romalis and Dean Karlin, for instance, opted for their own enhanced version of Weight Watchers. In their pact to stay trim, either of them can call an impromptu weigh-in, with the fine for weight gain being $10,000.[24] Karlin later teamed up with a different economics professor, Ian Ayres, to create *stickK.com,* a website to help others devise their own precommitment contracts. A similar but earlier effort is the website "Covenant Eyes," founded by Ronald DeHass. To curtail pornography consumption, it tracks and e-mails all your Internet visits to the "accountability partner" of your choice. It could be a friend, a spouse, or perhaps a pastor. For a technological solution in the same cast as Clocky, there is the alarm clock SnuzNLuz. Every time you press the snooze button, it donates ten or more dollars to your most detested charity; a little extra sleep comes at the cost of assisting groups that represent the antithesis of your political position, sexual orientation, or environmental stance.

Like all precommitment methods, these devices aren't foolproof. To begin with, they are inflexible, so you can't change your mind even for legitimate reasons. Where would Ulysses be if his ship started sinking or was attacked by pirates with him still bound to the mast? You might desperately need the money you've tied up in Christmas Clubs or fall ill and be unable to use long-term gym memberships. On the other hand, if you don't make them strong enough, disincentives can be circumvented. In keeping with the saying, "Those who flee temptation generally leave a forwarding address,"[25] be careful that your future self isn't smarter or more determined than your present version. If there is a will—and there most definitely *is* a will—then there'd better not be a way. Adults who nail-bite will coat their fingers with the same bitter ointment used to discourage children from sucking their thumbs, only to endure the taste or find inventive ways to wash it off.[26] Similarly, in Mordecai Richler's novel *Joshua Then and Now,* Joshua Shapiro helps his friend Seymour overcome a precommitment strategy by swapping underwear with him: Seymour was wearing "black satin panties with a delicate lace trim" to prevent his adulterous ways.[27] After all, what type of woman would want to sleep with you after she found out that you clad your manliness in lacey undergarments? Well, I guess it depends on what crowd you hang with, but that's beside the point.

1. Action Points for Commit Now to Bondage, Satiation, and Poison: Staying true to your goals can be a time-limited offer, requiring you to act before temptation overcomes you. You first need to identify your temptations, what distracts you when you should be working. If you need assistance, ask your family and friends. They likely know. After identifying your temptations, you have three options about what to do about them. Take your pick.

- *Bondage:* Put these temptations out of reach or at least far away. For example, erase your video games or disconnect your Internet connection. Remove the battery from your cell phone or unplug your television set.
- *Satiation:* Satisfy your needs before they get too intense and distract you from your work. Ironically, you can often work harder if you first schedule in some time for leisure.
- *Poison:* Add disincentives to your temptations to make them sufficiently unattractive. For example, a monetary bet with

someone else that you won't give in to your temptation can be applied to almost anything.

MAKING PAYING ATTENTION PAY

About the time I was born, the award-winning psychologist Walter Mischel started experimenting on children using marshmallows to test the power of their will.[28] In a series of studies, he would offer the kids a marshmallow, but tell them if they could wait a little while, they would get two marshmallows. Some waited a little, others a lot, with the average being about five minutes. The children's ability to delay gratification and get the larger but later treat proved critical as they grew up. The self-control they exhibited as kids predicted everything from their Scholastic Aptitude Test (SAT) scores to their adult social skills.[29] Character is indeed destiny. Subsequently, Mischel tried to change the destiny of a new cohort of children by improving their strategies for dealing with temptations, usually tripling their self-control, getting them to wait three times as long. What was his magic? He simply showed them how to pay attention.

Mischel's approach to conquering inattention will seem very familiar. As I do for the Procrastination Equation, Mischel emphasizes our mind's dual nature; procrastination arises from the interplay between our limbic system and our prefrontal cortex. To master attentional control as a means of increasing our self-control, we must first go from the inside out, to change what we see and how we see the world. Second, we go from the outside in, to remove or reinforce external cues, changing the world we see.

INSIDE OUT: PAY ATTENTION PLEASE!

It is time to play a game called "The Unlikely Beast." It will take precisely a minute. Take out your watch to time yourself and for the entire minute don't think of a *pink elephant.* No pink elephants, not even one. Got it? Since you probably didn't think of any pink elephants today, this should be pretty easy. If you can make it an additional sixty seconds without thinking of pink elephants, you win. Are you ready? Go!

INSERT SIXTY SECONDS HERE

Did you win? I doubt it. According to Daniel Wegner, who wrote the book on thought suppression, the game is rigged against you.[30] To make sure you aren't thinking about pink elephants, you have to keep some notion of them in mind, otherwise you can't watch out for violations. Ironically, by actively suppressing thoughts, you help to maintain them. This mechanism forms the basis of Freudian slips; trying to repress a trauma or a temptation seems to cause the dreaded idea to surface. For the few of you who did suppress the beast successfully for a whole sixty seconds, did you notice the post-suppression rebound? Your mind, in a sigh of release, probably indulged in a series of pink elephant fantasies as soon as the time was up.[31] Despite its disastrous track record, thought suppression is a popular technique used to combat—ineffectively—everything from homosexual urges to racial stereotypes. If you find yourself pestered with an intrusive temptation, whether it be for an illicit lover or a new television show, you can find better ways to stop thinking about it. Here's what works.

Instead of avoiding thinking about your temptation, you can mentally distance yourself from it by framing your temptation in terms of its abstract and symbolic features. For example, Mischel had children delay eating pretzels by having them focus on the snack's shape and color ("the pretzels are long and thin like little logs") rather than on their taste and texture.[32] Similarly, anthropologist Terrence Deacon managed to get chimpanzees to make food choices more strategically by using a form of symbolic representation called lexigrams.[33] The chimps were to choose between two portions of fruit, kiwis and strawberries, and received the fruit *they didn't select*. Only chimps who learned the lexigram equivalents of kiwis and strawberries (respectively a black square with a blue "Ki" versus a red square with two horizontal white lines) were able to enact the winning strategy of pointing to the less desirable fruit option and, in return, receive the more desirable one. As Deacon concluded, seeing the world in symbols tips the balance away from the stimuli-driven limbic system toward the abstraction-loving prefrontal cortex, enabling us to make better choices.* To take advantage of this quirk, we need to keep our thoughts as airy and formless as possible, as if seeing temptations from a great distance. As the seventeenth-century Japanese swordsman Miyamoto Musashi wrote in the *Book of Five Rings:* "Perception is strong and sight weak. In strategy it is important to see distant things as if they were close and to take a distanced view of close things."

Your second line of defense is to run a "smear campaign" on whatever features your limbic system finds desirable. You can ascribe negative qualities and consequences to every temptation to counteract its enticing features. Those pretzels, for example, could be stale or sneezed upon. The more such disgusting possibilities you generate, the more unpleasant the indulgence will seem.[34] Furthermore, by imagining some really horrific outcomes,

* Or, in Deacon's own words, chimps need the assistance of symbolic representation, for without it "being completely focused on what they want, they seem unable to stand back from the situation, so to speak, and subjugate their desire to the pragmatic context."

you engage in something called *covert sensitization.*[35] This technique is to pair your temptation with an undesirable image, hopefully infusing the former with the latter. Here is a generic one I developed specifically for procrastination:

> I want you to imagine you've just put off a major project, one that you thought you still had plenty of time for. You are doing other less important work, surfing the Internet, watching TV at home—procrastinating. Finally, the moment comes when you can't really put it off any longer and, though it will be stressful, you should be able to handle it—except you just came down with a throbbing headache. Given all the extra time you had to take on the project, you can't use this as an excuse without looking lazy and incompetent. You start working on it, but the headache gets worse and worse, like a knife twisting behind your eyes. You are producing nothing of value despite the excruciating pain as you try to work. As your eyes almost tear up with agony, you take some pain medication only to find that it makes you sleepy, and indeed you do sleep. When you wake up, it is morning and you are late for work. Rushing to get there, you find that your boss has decided to gather all of your colleagues in the boardroom for you to present your project. The president of your company stops by and decides to listen in too. Being late, you are rushed to the front of the podium and everyone waits for you to get started. As you try to explain you have accomplished nothing because of a headache, you stumble over your words and look like an utter fool. There is a long silence broken only by a few sniggers, with your colleagues looking away, embarrassed to be associated with you. Afterward, your boss explains that she was

> thinking of promoting you but now she will have to fire you instead—what you have done is inexcusable. One of the people at the meeting recorded your "presentation" with her cell phone and posted it on YouTube, where people everywhere mock you. No one in your industry will even give you an interview and your career is ruined.[36]

Feel free to change this scenario to fit your situation, tailoring it to your specific distractions. Joshua Shapiro's friend Seymour, for example, might have had better luck with fidelity by focusing on negative possibilities, like getting a stranger pregnant, catching a disease, or destroying his marriage. For yourself, just remember that when you leave tasks to the last moment, you can get sick, competing emergencies do happen, and work almost always takes longer than you thought. As for the dire outcomes that result from your procrastination, imagine the worst. The consulting company Opera Solutions lost a million-dollar contest by submitting their solution twenty minutes too late.[37] Elisha Gray lost credit for inventing the telephone to Alexander Graham Bell by submitting his idea to the patent office a day late. Delay makes bad things happen. Why not to you?

Attentional control and covert sensitization aren't perfect techniques, though. They require effort, and will eventually exhaust your energy stores—you can't avert your eyes forever. As Mischel's work showed, children's ability to delay gratification was increased, but remained limited. Still, some delay may be enough for your purposes. Many temptations are time sensitive, like dessert at the end of dinner; if you can avoid them for an hour or so, the desire to indulge will disappear. It isn't perfect, but it is better. If you are looking for more long-lasting solutions, read on.[38]

OUTSIDE IN: NOW YOU SEE IT, NOW YOU DON'T

Here is a trick that will give you an extra month of efficiency each year. It is easy to implement, immediately effective, and doesn't cost a cent. First, go to your e-mail program. Second, disable all the audio alerts and mailbox pop-ups. In Microsoft Outlook, they are buried pretty deep under "Advanced E-Mail Options," but the controls are definitely there. Just unclick everything under "When new items arrive in my Inbox." That's it, there is no third step. Banishing e-mail notifications will make you about 10 percent more efficient and over a year that translates into one more month of productivity.* The best work happens when you engage deeply in a single task. Every time you stop your flow, you have to once again decide to work and then it takes time to become fully re-engaged. Unfortunately, we are conditioned to answer e-mail instantly, responding to the tell-tale "ding" like Pavlov's dogs. Unless you have a pressing reason, check your e-mail at your convenience, during natural breaks in your productivity.

What we are doing here by changing our e-mail settings is regaining *stimulus control.* Part of our decision making occurs subconsciously, in our limbic system. This is not the brightest part of our minds; it takes much of its lead from environmental cues—that is, from the *stimuli* of sight, smell, sound or touch.[39] A provocative image pops up and we think of sex, a tasty smell wafts our way and we become hungry, or we hear a snippet of a song and start humming the tune. These associative cues cause our mind to wander and we forget the original task. With just a little nudge, our imagination slips down the rabbit-hole and we find ourselves mulling over some more personally relevant issue, like what's for lunch. We have been distracted.

---

* At least a month. See chapter 5.

These distracting cues are powerful and pervasive, and are actively pumped into our world. John Bargh, head of Yale's Automaticity in Cognition, Motivation and Emotion (ACME) Lab, has spent decades showing how little it takes to influence our minds.[40] We can be prepared—primed—for almost anything, all without being aware of it.[41] A slight dimming of the lights increases our fearfulness. Hold a hot cup of coffee and warm feelings infuse us, causing us to be more charitable. Putting Hershey's chocolate kisses on a secretary's desk in a clear rather than opaque bowl, thereby making them more visible but not more available, increases snacking at the office by 46 percent.[42] The power of cues is such that they can create cravings that leap upon us—"If you speak of the Devil, so he will appear." Addicts often feel an overwhelming urge to relapse when they encounter a strong drug cue, such as a neighborhood hangout or a former fellow user.[43]

Big business has been aggressively trying to direct these cognitive cues, deluging us with over a thousand advertisements each day. To take back control of our environment, essentially we need to run our own personal advertising department. As it is, our workplaces and schools are motivationally toxic, polluted with distractions. We need to make them sanctuaries of performance, taking advantage of the "out of sight, out of mind" adage to purge our offices and classrooms of irrelevant cues. At the beginning of this section, I asked you to turn off all your e-mail alerts. I also told you about how Ulysses had his crew seal their ears with wax to avoid hearing the Sirens. Both of these examples draw upon the same principle of eliminating external cues. You need to identify your distractions and cleanse their accompanying cues from your life. I bet you have more than a few Internet sites hot-linked on your computer for easy access. Start by deleting those. While you are at it, get rid of any quick-launch icons for games, or better yet, erase the games completely. At home, hide the remote control for the TV or close the doors of the television cabinet if you have one. Now for the really hard part.

A messy workspace, cluttered and disorganized, is a minefield of distractions. For every minute you hunt for a misplaced report or book, the likelihood increases that some tangential tidbit will entrance you. Everything extraneous on your desk distracts and detracts, making it harder to find and focus on your primary purpose.[44] But here is the catch-22: the number one activity that people postpone is "cleaning out closets, drawers, and other cluttered spaces."[45] Procrastinators are more likely to leave clutter, which in turn, increases their procrastination.[46] You need some help. You can combat clutter with some of the other procrastination-fighting techniques in this book—the structured or productive procrastination we looked at in chapter 8 is particularly relevant for you. The most motivating time to de-clutter your life always seems to be before another pressing deadline. Alternatively, look outside these pages for help. Just search online under the word "clutter" to find books on how to organize your life. You can also call in organizing experts; it's no more unusual than hiring a personal trainer to jumpstart your exercise program.*

Once you have banished the signs of temptation, the other half of this stimulus-control strategy is filling the void. External reminders of our goals are important, but instead of motivational posters featuring generic catch phrases, your reminders need to be personally relevant. They need to speak to *you.* What do you strongly associate with the target task? If there is

* Here are two national associations: http://www.napo.net/about_napo/; http://www.organizersincanada.com/.

a quotation you find particularly inspiring, have your screensaver produce it whenever you idle. If you are slow at paying bills or taxes, place them prominently on your kitchen or coffee table, where you can't ignore them. Even writing a list is a good reminder, especially on a sticky note posted to the side of your computer screen.[47] All these cues solidify into an unwaveringly effective concentrative strategy, focusing your attention toward your goal.[48]

To emphasize how effective this concentrative strategy can be, consider the boost it can give to your household energy efficiency. The problem with energy consumption is that it is distant and vague, only realized in a monthly bill long after the kilowatts have been killed. If we made a very small change and put your electricity meter on the *inside* rather than the *outside* of your house, this visible and constant reminder of your energy cost would coordinate your limbic system with your prefrontal cortex, sparking you to turn off unneeded lights and replace the remainder with efficient fluorescents.[49] Mark Martinez from Southern California Edison, for example, had his customers use an *Ambient Orb* that glowed red when electricity was expensive.[50] Within weeks, peak hour consumption voluntarily reduced by 40 percent; and other similar experiments have indicated about a 10 percent savings in monthly utility bills.[51]

For work, stimulus cues don't have to be store-bought. Anything associated with a task can spur you to complete it: time of day, preceding activity, and colleagues all can be transformed into work triggers.[52] Most usefully, you can make your place of work itself a cue, so that focus comes automatically as soon as you sit down. This strategy requires dedicating your environment exclusively to labor. To do this, work in your office until your motivation leaves you and goofing off becomes irresistible. At this point, do your web surfing, your social networking, your game playing *somewhere else*. This may require you to get a second computer, one for play, but when the added productivity kicks in, the purchase will pay for itself. If you keep work and play in discrete domains, associations will build and attention will become effortless—your environment will be doing all the heavy motivational lifting. Three studies have investigated the effectiveness of this technique with students, and found that the use of dedicated work areas decreased procrastination significantly within weeks.[53] Similar applications, such as using separate banking accounts to prevent impulsive spending, can be almost instantaneously effective.[54] Without this segregation between work and play, you get conflicting cues every time you sit down at your desk, one indicating that you should research your report and the other egging you on to check your Facebook page.

To sharpen role boundaries between clashing life domains, typically family and work, we need to keep the demarcation lines pristine.[55] If you can't afford a separate computer, then at least create a second profile that requires you to log out of your workplace identity before you slip into your lazier alter ego. If you find your BlackBerry allows the office to pollute family time, get a stripped-down second cell phone to use when you punch out. You might also include a transition ritual to help you move from one domain to another, such as winding down with the radio during your commute or changing out of your "work clothes" when you arrive home. If you need to work at home, have a separate office, no matter how small or symbolic. These environmental cues will fence off distracting temptations, allowing you to truly *be* in each place.

2. *Action Points for Making Paying Attention Pay:* Distractions are a major enabler of procrastination, so learning how to effectively handle them is a must. Your options are to denigrate, eliminate, or replace cues that remind you of your temptations.

- Sully tempting alternatives by using covert sensitization, imagining disgusting ways they may be tainted, or envision possible disastrous outcomes from procrastinating. The more vividly you can imagine the contamination or the catastrophe, the more effective this technique will be.
- When confronted with distracting temptations, focus on their most abstract aspects. Triple chocolate cheesecake, for example, can be construed as another fat and sugar combination.
- Entirely eliminate cues that remind you of distracting alternatives where possible. Keeping your workplace clear of clutter will help you accomplish this.
- Once you have purged your workplace of distracting cues, replace them with meaningful messages or pictures that remind you of why you are working. For some, a desk photo of loved ones can be an effective reminder.
- Foster these work cues by compartmentalizing your place of work and play, keeping them as separate as possible.

SCORING GOALS

Inch by inch, life's a cinch; yard by yard, life is hard. How powerful is this mantra? Joe Simpson, in one of mountaineering's greatest survival stories, used it to save his life. Left for dead at the bottom of a crevasse in an isolated Peruvian mountain with a shattered shinbone, he had three days to pull himself to a base camp through five miles of truly treacherous glacier field or be really dead. He was already utterly exhausted from an arduous marathon of an ascent, with no food and only a little water, so this journey should have been impossible, except for one critical survival tool: his wristwatch. With it, he set goals. Setting the alarm for twenty minutes at a time, he made for a nearby rock or drift—he was elated when he reached it in time and he despaired when he didn't. Battling exhaustion, pain, and eventually delirium, he repeated this process hundreds of times and reached the perimeter of the base camp just hours before his friends' intended departure.

Simpson's story, recounted in his book *Touching the Void,* highlights the power of goal setting. As Mark Twain wrote: "The secret of getting ahead is getting started. The secret of getting started is breaking your complex overwhelming tasks into small manageable tasks, and then starting on the first one." Further notions about how to construct goals to maximize their motivational benefits, however, are shrouded in confusion. Despite thousands of scientific studies on how best to set goals, little of this know-how has permeated into the mainstream.[56] Since the mid-eighties, over five hundred books have stressed S.M.A.R.T. goals, an acronym that has both too many and too few letters. S.M.A.R.T. stands for: Specific, Measurable, Attainable, Realistic, and Time-Anchored. There are too many letters, in that *Specific* is redundant with both *Measurable* and *Time-Anchored* while *Attainable* is redundant with *Realistic.*[57] There are too few in that it is still missing major concepts. Let me tell you what you actually need to know.

We have already touched on some of what makes a goal good. In chapter 7, we mentioned that making goals challenging is more inspiring than making them attainable. Easy goals are attainable. You know what happens after obtaining your easy goal? The same thing that happens after you cross the finish line of any race: you stop.[58] In chapter 8, we focused on making goals meaningful by linking them to personally relevant aspirations.[59] If you see how present tasks lead to future rewards, you will value them more highly. In this chapter, we will put the finishing touches on goal setting by putting time back on your side.

THE FINISH LINE IS JUST AHEAD

Almost invariably, reporters contact me about their piece on procrastination mere hours before it is due. *Slate* magazine, for example, which did a special issue on procrastination, confessed: it was "originally planned for the week of May 5. Seriously. We'd planned to publish that Monday morning, but there was one problem: only a handful of our writers had managed to get their work in on time."[60] My theory is that the fourth estate is full of unrepentant procrastinators, drawn there because it is one of the few places they fit. Every day the job itself generates a specific and proximal deadline: so many words on this topic by this hour *or else!* This is exactly the type of goal that procrastinators excel at meeting. To get motivated, they need a clear and close finish line. Their action curve follows directly from the Procrastination Equation; as delay shrinks, motivation peaks.

To apply this principle to your life, you need a concrete and exact notion of what needs to be done because vague and abstract goals (such as "Do your best!") rarely lead to anything excellent. The level of detail required differs from person to person but you should be able to sense when you've got enough. Goals should have a corporeal rather than an ethereal feel—you should be able to sink your teeth into them. "Complete my Last Will and Testament before flying on the 15th" is an achievable goal. "Get my finances together," not so much.

After creating a specific finish line, schedule it soon. You may need to break up a long-term project into a series of smaller steps. Consider the following chart, which represents most work situations. In the background, there is always a buzz of temptation and though it will have its peaks and valleys, on average, we can represent it by a straight horizontal dashed line. Until our desire for work exceeds this constant, we won't be working. Typically, we allow the environment to set our goals for us and it is pictured by a single goal: the deadline. The triangle line represents a person with no self-set goals, whose motivation is mostly reserved until just before the deadline. What to do? How about artificially moving the deadline closer? The unadorned solid line represents a person who has broken down the task into two earlier subgoals, allowing work motivation to crest above the temptation line sooner. As can be seen, the sum of the parts can be greater than the whole, as the person who sets subgoals works for twice as long as the person who doesn't.

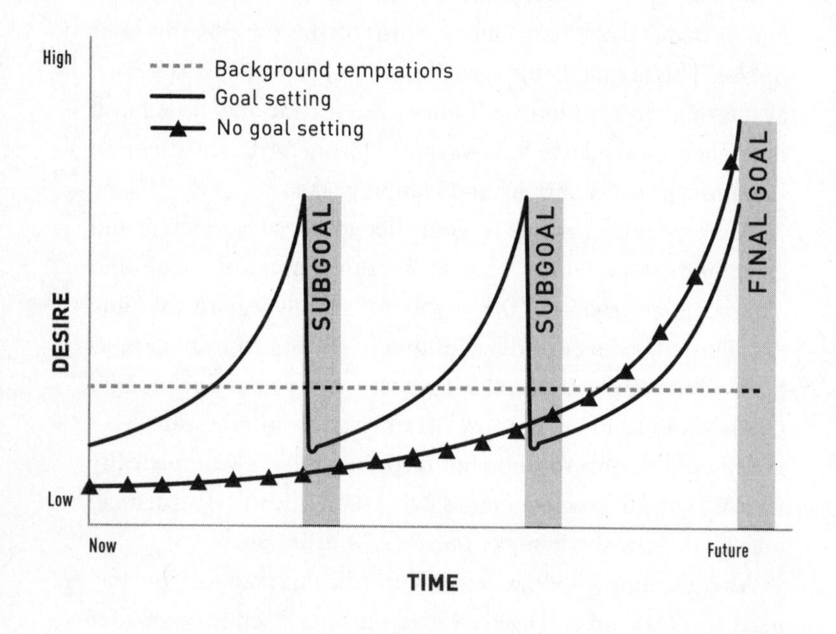

There are no hard rules for how specific and how proximal your goals must be to be effective. Your success depends on how impulsive you are, how unappealing you find the task, and what temptations you are battling. But keep in mind that too-frequent goals can be cumbersome. Daily goals typically provide a good balance; they are both effective and practical. Still, many find that the hard outer shell of a chore, the first few minutes, remains the initial obstacle. How many times have you put off a task only to realize it wasn't so bad once you got started? Cleaning, exercising, and even writing are often difficult at first. It is a bit like swimming in the lake by my in-laws' cabin, just north-east of Winnipeg (the coldest city in the world with a population greater than 600,000). The water is deliciously invigorating but, for most, the initial temperature shock is an effective barrier against reaping the subsequent reward. By focusing solely on the initial jump off the dock, I can plunge in and, after a few intense seconds, enjoy myself. An extremely short-term or mini-goal, then, is excellent for busting through such motivational surface tension. Ten-minute goals are an application of this technique, such as the ten-minute clean-up around the house. Consequently, if you have trouble writing, just sit down and type a few words. If you don't want to exercise, at least get your workout clothes on and drive to the gym. Once you have completed your mini-goal, re-evaluate how you feel and see if you are willing to immediately commit to a longer stretch. Having broken through that motivational surface tension and immersed yourself in the project, you, like most, will opt to continue.

Your final choice is how to structure your goals. Do you prefer *inputs,* the time invested, or *outputs,* what is produced? For exercise, are you going to run for an hour or for five miles? Both are good options. A modest but regular schedule, if it really is regular, produces wonders. B.F. Skinner thought "fifteen minutes [of writing] a day, every day, adds up to about a book every year," though most professional writers aim to do far more than a quarter of an hour.[61] Others go by the word count; science fiction author Robert Sawyer, for one, writes two thousand words each day, including his blog. Ernest Hemingway combined both inputs and outputs, writing for six hours or producing about five hundred words, a useful strategy. If you have a fruitful day and hit your output quota early, be it words or widgets, reward yourself and go fishing; if the productivity doesn't come, the input or time requirement ensures that something is produced. To help keep you honest about your productivity, try using free software like *ManicTime* or *RescueTime*.[62] They are nifty applications that automatically track your computer work habits, allowing you to easily monitor your activities. How much time are you spending on e-mail? How about web surfing? How much do you actually spend on work? This kind of reality check will make you aware of your productivity and I'll personally vouch that it is useful for winding down an Internet gaming habit.

FULL AUTO

Occasionally, on my ride home from work, I am charged with stopping off at the grocery store to pick up milk or diapers. This side trip entails taking the earlier exit off the highway, which I invariably drive right past. I then need to negotiate a laborious series of travel corrections to get to where I should already be. The problem is that I've done my commute so many times that I'm on autopilot. We have dozens of these automatic routines in our lives, which we can perform even when dead tired. In a mindless blur, we eat breakfast, brush our teeth, and tie our shoes. Despite their zombie-like quality, these routines have power we can tap—the force of habit.

Both the strength and weakness of routines lies in their lack of flexibility. Their weakness is that once we fall into a habit, we tend to follow through even when a change of pace would be beneficial. We go to the same restaurants, order the same food, watch the same shows, without really considering possibly better options.[63] On the other hand, routines are easy to maintain and can be undertaken even when we're exhausted.[64] By intentionally adopting a routine, we can pursue long-term goals even when our wills are weary and temptations abound. We push forward oblivious to other choices, choices that might mean stopping, resting, doing otherwise. The fewer moments of choice there are, the less likely you will be to procrastinate.[65] That is, if you have the right kind of habits. Routines are like Don Quixote's windmills; they can raise you up to the heavens or drop you down into the mud. Though we have our share of bad habits—reflexively turning on the TV or finishing a bag of potato chips—we can create good ones. We can turn exercising, cleaning, or working into at least semi-automatic routines. Scientific study confirms the benefit of this effort; procrastinators perform as well as anyone else when the work is routine.[66]

Building a routine requires activating many of the same precepts as stimulus cues. You want *predictability*. Devise rituals of performance, keeping as many of the environmental variables as stable as possible, especially time and place.[67] Exercise programs, for example, should take place at regularly scheduled times, leaving little guesswork about where and what the fitness activities will be. Like clockwork, every Tuesday afternoon at 5:00, you go and lift weights, and every Thursday morning at 6:00, you go running. Take whatever you have been putting off and specify where and how you intend to implement it. For instance, make a vow: "When breakfast is finished on Saturday morning, I will clean out the storage room." This seems so easy and simple that it couldn't work, but it does. When you make an explicit intention to act, the desired behavior just happens.

The expert on the psychology of intentions, Peter Gollwitzer, finds that forming intentions almost doubles the chances that you will follow through with almost any activity. The effectiveness of explicit intentions has been scientifically confirmed on everything from cervical screening to testicular self-examinations and from recycling to writing a research report over the holidays.[68] In terms of ease and power, this is as good as it gets. Making an intention is a remarkably accessible back door into your brain; it programs your limbic system to effortlessly act on cue as you see fit. Intentions can even be used to implement other self-regulatory techniques, especially when expressed in an "If . . . then" format. If you have energy issues, make the intention of "*If* I get tired, *then* I will persevere." If you are easily distracted, it would be "*If* I lose focus, *then* I will move my attention back to the task." And of course, "*If* I am pursuing a goal, *then* I will use implementation intentions."

Be warned that when trying to start your routine, you will invent a ceaseless onslaught of excuses not to follow through. You will get sick, go on vacation, have extra work, fall behind elsewhere, and find it ever so convenient to let your schedule slip. Defend fiercely against these slippages! Routines get stronger with repetition, so every time you slack off, you weaken your habit and it becomes even harder to follow through the next time. If you protect your routine, eventually it will protect you.[69] At the start, your regimen will need constant nursing.[70] Some temporary professional assistance can be a good investment; after all, you are investing in yourself. Personal trainers to run you through your paces or professional organizers to help you clean up can help launch you in the right direction.[71] To draft your last will and testament, hire an estate planner or a wills and estates lawyer.[72] They provide as much motivational help as legal expertise, structuring the process to maximize your follow-through. But hired help can't do it on their own, nor can this or any other book. In the end, the responsibility lies where it has always been—with you.

3. Action Points for Scoring Goals: This is really saving the best for last. Goal setting—proper goal setting—is the *smartest* thing you can do to battle procrastination. Though every other technique discussed so far has its place, goal setting alone may be all you need. Along with making your goals challenging (chapter 7) and meaningful (chapter 8), follow these remaining steps. Regardless of what other books say, this is what's proven to maximize your motivation.

- Frame your goals in specific terms so that you know precisely when you have to achieve them. What exactly do you have to do? And when do you have to do it by? Instead of "Do my expense report" it should be "Gather all my receipts, itemize them and record them by lunchtime tomorrow."
- Break down long-term goals into a series of short-term objectives. For particularly daunting tasks, begin with a mini-goal to break the motivational surface tension. For example, a goal of tackling just the first few pages of any required reading can often be enough to get you to finish the entire text.
- Organize your goals into routines that occur regularly at the same time and place. Predictability is your pal, so open your schedule and pencil in reoccurring tasks. Better yet, use an indelible pen.

LOOKING FORWARD

If only time-sensitive Tom could have read this chapter! He put off booking his hotel and subsequently had a vacation to forget instead of one to remember. He probably didn't even need all the techniques in this chapter to have changed his fate.

Perhaps it would have been enough to set a specific deadline for himself, say, next Thursday night, and frame his intention to act in explicit terms, as in: "Immediately after dinner I will research hotels in the area and book a room." For good measure, he could have imagined some worst-case scenarios, such as: if he continued to procrastinate, then his room would be far away from the beach and in desperate need of redecoration. Those of you who scored 24 and above on the impulsiveness self-assessment scale in chapter 2 should pay special attention to the techniques here, but almost everyone would benefit from them as well. Though some of us are more impulsive than others, we all can make regrettably impulsive choices.

The fundamental challenge in implementing these steps is that attempts to increase self-control require some self-control to begin with. The obstacle is similar to strength training; in order to initiate the process, we need to be able to lift at least the lightest of the available weights. As for procrastination, the worse it is, the harder it becomes to remedy. The very motivational deficits that create your procrastination also hamper your attempts at change. If you are unable to delay gratification, for example, methods to increase your patience must initially be immediately rewarding in themselves. Otherwise, advice becomes useless shouting from the sidelines, annoyingly extolling you to "do first things first." If you could simply do that, you wouldn't need the advice in the first place. Fortunately, most of these techniques are easy to adopt, like turning off your e-mail ding or making those explicit intentions to act. These immediate successes will give you the confidence and the self-control to increase your efforts, all of which will become even easier with practice. From here on out, life becomes better, not harder.

CHAPTER TEN

Making It Work

PUTTING THE PIECES INTO PRACTICE

Do or do not do. There is no try.

MASTER YODA

Before I get into this chapter, I want to thank you for persevering. People who procrastinate tend to get distracted and turn to other things. So since you have reached chapter 10—and I am assuming you haven't skipped ahead to the end—you deserve a little praise. After all, the tendency to put off has such a deep resonance in our beings that it is more remarkable when we *don't* procrastinate than when we do. Having read through the book, you have a good grasp of the underpinnings of procrastination, how it emerges from our brain's architecture, the ways in which the modern world makes it worse, and what you can do about diminishing it. There is just one last step to putting procrastination in its place. You need to believe what you read.

I can't really blame you if you are a little suspicious. If you are familiar with self-help books, you have certainly earned some cynicism. There is so much misinformation in the field of motivation—so many promises that don't deliver—that "What if someone wrote a self-help book that actually worked?" is the premise of Will Ferguson's international bestselling novel *Happiness™*. Satirizing the self-help industry, Ferguson invents the character Tupak Soiree, who writes *What I Learned on the Mountain,* a tome that genuinely helps you lose weight, make money, be happy, and have great sex.* Now I can't promise the last of these, but *The Procrastination Equation* is about making the rest of *What I Learned on the Mountain* a reality. Every technique in this book is based on the bedrock of scientific study, so it had better work. Just flip ahead a few more pages and look at the research I have laid out in the Endnotes.

The Procrastination Equation, just like *What I Learned on the Mountain,* is still only an inconsequential book if the techniques stay locked inside its covers. In Ferguson's novel, the challenge was just getting people to read it. For a while, *What I Learned on the Mountain*'s potential effectiveness was derailed, as you might guess, by procrastination. As Edwin, the book's editor, concludes: "I forgot about procrastinators. Don't you see? All those people out there who purchased the book or were given it as a gift and still haven't got around to reading it." For my book, the requirements are a little steeper, but as you can see, you have already pretty much finished it. To make what you are reading effective, you also need to take its contents seriously. You need to adopt these techniques into your life and start seeing your decision making in terms of that interplay between your limbic system and your prefrontal cortex. To lift the ideas off these pages and into your life, we are going to take one parting look at Eddie, Valerie, and Tom and imagine how

---

* More or less. I don't want to spoil his plot.

they are getting along. You'll see that they are using all these techniques in combination, and thriving because of it. And if you can see yourself doing the same, then you will be able to get your act together, and you will soon be putting procrastination behind you as well.

EDDIE AND VALERIE

After Eddie lost his sales job, he was depressed for a long time—that is, until he met Valerie. She always found a way of putting a smile on his face and it was natural that the two got married. Now in their thirties, with two full-time jobs and a lovable toddler named Constance, they have a wonderful life. But they are always on the run, and lately the demands have been getting worse.

Valerie is often on crushing deadlines, and her home responsibilities take second place when she is in a crunch. She knows how lucky she is to have a job at the local newspaper, but there have been cuts, and she is now doing the work of two people, maybe more. The pressure to meet all her deadlines is serious—this isn't about career advancement, it's about staying employed. Eddie has to travel for his job in marketing, which means that he leaves before dawn and is away for days, leaving Valerie in a lurch. When Constance gets sick, all hell breaks loose. She keeps them up at night, and somebody has to stay home with her. When the washing machine breaks down, somebody has to wait for the repairman. Valerie and Eddie feel as if they haven't had enough sleep in years. And they are right. They know how lucky they are to have two jobs and their little girl, but they are stressed beyond words.

Valerie and Eddie shuttle between work and home like mechanical dolls, always late, grabbing a kiss or a donut on their way out. When they are at home, they worry about the work they are not doing, and so they often go to the computer after the baby is asleep, working through exhaustion late into the night. If the baby is sick, the one who goes to work frets about how she is, and when she is well, they are both checking her out on the webcam at daycare—spending precious work minutes monitoring her well-being. They can hardly handle paying the bills and getting to the pediatrician's office for checkups and shots. They e-mail each other dozens of times a day, and Eddie has to control himself from texting Valerie from the car on the way to his next meeting.

Eddie promised himself he would clean out the garage last summer, but it's October, and the junk remains. Valerie has lost control of her vegetable garden, which she started as an altruistic family project but which has devolved into a sad collection of wilted greenery. They are considering canceling their joint membership in the gym—they are both too tired to work out at the end of the day, and mornings reach a level of chaos that drives them both nuts—dressing the baby, exchanging directives about multiple tasks, suddenly full diapers and fussy moments . . . you can fill in the blanks.

This is actually the best-case scenario. It could easily be worse. They face no sudden illness, no job loss, no financial straits, and no tragedy. But Eddie and Valerie's lives are out of control and they are facing the conflicts that every working couple with kids has to deal with. Recently, Valerie began to feel that she is never in the right place—at work, she thinks she should be home; and at home, she worries about all the work she should be doing. She is feeling frayed and tattered, and is starting to hate her life. Looking for some cheering up, she calls her sister, who listens sympathetically, and then offers a little advice: "There's this book I've been reading that has a few ideas that might help. Do you want to borrow it?"

Like all such offered books, this one was gratefully accepted but put aside. That is until one stressful sleepless night, when Valerie in desperation decided to crack it open. After skimming through the pages, she noticed the research behind it. "Well now," she thought. "This stuff has really been battle-tested. Let's see what I can find for Eddie and me." Taking some paper and a pencil, she slowed down and made some notes about what she might be able to use.

The next night when Eddie shuffled home, Valerie sat him down and told him flat out, "I'm not happy. Things are going to have to change."

Eddie sighed and, revealing his low expectancy, said, "I'm not happy either, but this is just the way life is. We can't change it."

"You always say that and you're usually wrong," Valerie replied. "I think there are steps we can take to make our life better. My sister lent me a book and it's based on scientific research. I hear it has helped a lot of people and we could use some help ourselves. I think we should at least try some of the ideas. For starters, we just need to lay down a few goals."

Eddie was too exhausted to argue with her, so he played along. "I have a goal," he said with a small smile, "I want to be happy."

"They have to be *specific* goals," Valerie said patiently. "They have to be concrete and doable, something we can get excited about."

"How about I want to be happy today?" Eddie suggested.

Valerie thumbed her way to the relevant page of the book. "We start by making some goals about the minimum changes we need to make to stay sane. I need to see my friends more often. I haven't seen them properly since Constance's baby shower and talking this over with them always makes me feel like my problems are more manageable."

Slumping into a chair, Eddie sullenly replied, "And my goal is to hit the gym every weeknight."

Valerie kept on message. "Get realistic. I think you can spare me one evening every other week. In return, I'm willing to cover you every Saturday morning if you want to exercise."

"That would be nice," admitted Eddie. "But I don't think I am up for handling an evening with Constance on my own."

Valerie pointed out that he often bathed Constance and put her to bed. "I want you to imagine hitting that gym, Eddie, how good your muscles are going to feel afterward. Also, imagine how much happier I'll be around here if I get some time with my friends. Can you picture that? Take a second and bask in its glow. Great! Now open your eyes and come back to reality. Does that give you the motivation?"

"All right," Eddie conceded, warming to the idea. "Let's do it."

With a little mental contrasting to spur them on, Valerie and Eddie's goal-setting techniques and "unschedule" (scheduling in realistic leisure time first) do indeed work. Valerie is seeing her friends, and after sharing her problems and hearing others deal effectively with their own issues, she is gaining a little more perspective. She is reassured that Constance will grow up and the economy will get better. It is amazing what a little social support (see *Vicarious Victory*) can do for a person. Eddie himself is glad to get to the gym once in a while. The exercise takes away a lot of his stress. He sleeps a little better and has more energy to tackle the rest of his life (see *Energy Crisis*). Still, a few weeks later, Eddie suddenly announces he has to work late and tells Valerie she has to cancel her plans. When he finally gets home, Valerie is not pleased.

Eddie pleads his case, "Look, I'm sorry you missed your night out but I had work to do and that takes precedence."

"Night out?" snapped Valerie. "It's more than a night out. I need that time with my friends. I wouldn't mind if you had to leave on one of your road trips for work but you e-mailed me fifteen times today while you were at the office."

"I thought you liked those texts!" retorted Eddie.

Composing herself, Valerie replied: "Here's what I like. I like face-to-face time with you and with my friends. For every minute you take to text me or send off an e-mail, that's ten minutes less we have at home. It takes ten minutes at least for you to get your mind back into your work after taking a break."

This surprised Eddie, but he wasn't going to give up his texting without a fight. "That may be so, but you text too. Besides, I can't work like a machine at the office. I need my breaks."

"And why are you tired?" asked Valerie.

"Well, it's impossible to get to bed early with all the evening work . . ." Then Eddie paused, making the connection. "Oh! Yeah, that might work."

"If we stop texting during business hours, stop Internet surfing, stop mindlessly checking our e-mails, that'll make at least two extra hours each day for the both of us. Hours we can use for sleep."

"My mind will zonk out from so much concentration," said Eddie.

"The book has a few ideas about how to make it work. Start with this. Create a second computer profile for yourself with a different background and layout. Log out of your regular work persona and into this play persona whenever you need a rest. If you aren't willing to take the minute to do it, you don't need the break. Here, I got you a present to help you commit."

"I like presents. What is it?"

Valerie pulled a silver-framed photo from her purse. "A framed picture of Constance and me. Every time you think of slacking off, this will remind you of why we're both pushing ourselves so hard. Remember, this is about us spending more time together as a family. Promise me you'll do this?"

"OK. I'll do it if you do," said Eddie.

And it works, of course. By ridding their workplace of their major temptations (see *Making Paying Attention Pay*), they have become more productive in the time they are at work and more relaxed when at home. They are starting to wind down for bed and are getting a better night's sleep, so that they can perform even better (see *Energy Crisis*). To help them get to where they need to be and remind them what this is all about, Eddie keeps that framed photo of his family on his desk (see *Games and Goals*), especially since it reminds him of what he really wants to do—spend more time at home, not texting at the office (setting approach goals, not avoidance goals). It didn't hurt that Valerie raised the stakes by extracting a little verbal precommitment from Eddie. In the end, they have a little more time than either expected, with both of them hitting the gym at least once or twice a week. Sicknesses, surprises, and other obligations still push them out of their routine, but now they are learning how to push back. They know they are fighting for a life that works. Eventually, Eddie even has the time to do some light reading, which he never used to have the energy for.

After putting Constance to bed, Eddie poured Valerie and himself a cup of tea and plopped into his comfy chair. "I've been looking through this book of yours," he said, "and I see where your ideas come from."

Picking up her own cup, Valerie replied, "Well, the secret was in actually following through with them not just reading the book."

"You're right," said Eddie, "but I have a suggestion of my own."

"Go on. I'm listening," said Valerie.

"Here's a technique called *Let Your Passion Be Your Vocation.*"

Her eyes widening in horror, Valerie gasped, "You're not thinking of leaving work to be a golf pro!"

"No, no, no, I'm not thinking that at all. Well maybe a little, but no," teased Eddie. "But how about this? Getting home earlier is reminding me of how much I used to love to cook. Remember those romantic meals I made for you when we first starting dating? Well, you don't mind cleaning up as much as I do. So, I'll tell you what: I'll do all the cooking if you do the cleaning."

Sweetening the arrangement, Valerie added, "If you throw in grocery shopping too, you've got a deal."

"If cleaning includes laundry, I'll shake on it," said Eddie.

"Done and done."

A sensible pair, they have now allocated the tasks of childrearing and housekeeping according to their differing tastes and talents. So Eddie does the cooking and shopping for groceries. He goes to the supermarket on Saturday or Sunday and stocks up for the week. This is easy for him because he loves shopping and the peace and quiet of chopping. Valerie, who never cared much about food, watches the baby when Eddie is doing the cooking. She cleans up after him, and she does the never-ending batches of laundry. Constance goes to daycare during the week, and they trade off taking her there early in the morning and picking her up after work. Life is getting better. Not insanely better. Not perfect. Just noticeably better. Valerie and Eddie are beginning to live life in harmony with who they are and what motivates them.

TIME-SENSITIVE TOM

On his journey back home from his disastrous vacation in the Dominican Republic, time-sensitive Tom was delayed at the airport for most of the day. It was hurricane season, which he had not thought about when he planned the trip. Sitting in the lounge, Tom reflected on his life. He was never much of a student, and constantly struggled with deadlines. But he knew that his friends at the fraternity were always glad to see him. An upbeat kind of guy, Tom always had a word of encouragement for the freshmen who were having trouble adjusting to college and being away from home for the first time; he enjoyed helping out. How did he get stuck in such a terrible rut? Without anything else to do, for hours he reflected on how much his procrastination had detracted from his own success, aspirations, and happiness. He thought about how it had affected not only his work life but also his home life. He realized that even if his vacation hadn't been such a mess, much of his leisure time would still have been focused on all the work waiting for him back at the office. He desperately yearned for that childhood feeling of unfettered time and guiltless play unpolluted by pressing obligations. His mind primed, he couldn't help but notice a title in one of the airport bookstores, a book that promised help. After buying it, he read it in its entirety during his wait and then on his flight home. Excited about the book's possibilities, he couldn't wait to put the techniques to use—this time his impulsiveness worked for him rather than against him.

On his first day back at work, Tom purged his office of temptations. He loaded software to keep track of his productivity, and he started setting specific, timely, and challenging goals. The results were immediate. Instead of being constantly behind, Tom found extra time to help others with their projects. "All the better," he thought; he always enjoyed talking and helping the people he worked with. Happy with the results, on a whim he went hardcore and used precommitment, promising to his boss that if he didn't get his next report finished in seven days, they could keep his upcoming year-end bonus. This got his boss's attention. When he handed the report in a day earlier than promised, people were amazed. What had happened to Tom in the Dominican Republic, they wondered. Over time Tom's interest in helping his colleagues and his fidelity to deadlines made his superiors think that he was showing leadership potential, and so they promoted him.

As the excitement of the promotion started to fade, Tom shared the news with his older brother Tim. After a few congratulatory drinks, Tom confessed it wasn't all good, "What did I get myself into? What do I know about leadership? I'm not a leader. I just barely learned how to get myself in shape. You know about these things. You took that leadership course back in college. What should I do?"*

Tim laughed, "Well, I guess it's too late to say 'don't panic.' But you have a right to be worried. No one who knew you a year ago would have expected you to be doing so well."

"Thanks for taking the pressure off, Tim," Tom replied sarcastically. "I guess you forgot all that leadership material anyway."

Rising to the bait, Tim put down his drink and focused. "Sorry. You're right; you do need to know this stuff. Leadership is important and not just for your organization's success. Most employees rate their relationship with their boss as their top concern. If you screw up, it can make your employees more miserable than

---

* Aside from being referenced in dozens of college textbooks, the Procrastination Equation is also used during managerial training programs. For example, the company Intulogy bases motivational training for managers around the Procrastination Equation and it works. As one of their clients testifies, "When you first told me that you wanted to introduce yet another motivation theory, I thought it was a big waste of my time. Yet, it worked in class. Then, I spent all summer thinking about the theory. I have realized how much it applies to everything in life. It's incredibly powerful."

if you took away a huge chunk of their paycheck. You now have the power to crush a considerable number of people's spirits."[1]

"And that's why I'm talking to you," said Tom.

"Well, I'm happy to help," Tim replied. "I've been thumbing through that book you lent me and most of the basic leadership techniques are already laid out—you just need to apply them to other people, just the way you did when you applied them to yourself. You can practice leadership along with self-leadership."

"Good, because I am not planning to go back to college," said Tom. "Let's get down to it."

Tim looked up at the ceiling, trying to remember the details. "There are two basic leadership styles: *transformational,* a people-oriented approach, and *transactional,* a task-oriented approach.[2] Since you're a people person, Tom, start using your people skills—go transformational!"

"So buddy up to them?" asked Tom.

"Nope," said Tim. "The first thing to do is to focus on creating confidence. What you need is an early success, to help them build faith in you and their ability to succeed under you. It's a basic principle, that you create achievable goals to recognize and celebrate. Later, this will help give everyone the confidence to persevere and hit the harder milestones."

"Ah, create a success spiral!" exclaimed Tom, making the connection.

"Exactly!" said Tim. "I knew a teacher who did this. She built confidence in us by starting off the semester with a few simple quizzes before proceeding into more difficult assignments. I really had a crush on her. One time after class, I remember . . ."

" . . . you're going off topic," interrupted Tom.

"Where was I?" said Tim, finishing his drink. "Well, you can also use the vicarious victory principle by setting the tone.

Confidently and clearly articulate a vision of where you want to be, exude optimism, provide pep-talks and in general be the role model. It's textbook."

"Me? Be the role model? What are you thinking?" Tom complained.

"Heavy is the crown . . . Of course, you could always quit or just take their money and wait for them to fire you. To me, that sounds a little bit like stealing, but I guess you have your own moral compass . . ."

Tim looked expectantly at Tom, letting the point linger.

"All right, all right, I'll do it," said Tom. "I was just thinking it through."

So on Tom's first day in charge, he gathered his staff together and gave them a prepared speech about what he intended to accomplish. He told them that though there were areas of excellence in what they had been doing, they were taking too long to finish financial reports despite logging tons of overtime. He then set that first achievable goal. "For starters," he enthusiastically told them, "I want us to cut the average time we take to compile our reports by a day this month. I think we can do it. In fact, I know we can do it." And Tom did know; it was a pretty easy goal. Still, he stayed on message at their weekly meeting, realizing that enthusiasm can be contagious. And at the end of the month, he found that indeed they did cut their production time by a day, precisely one day. "That's a start," he thought to himself, "but really we need to cut our production time by a week." He phoned Tim about his success and his situation.

"Well, that's great news," said Tim. "It's one thing for you to ask for advice but for you to actually follow through is impressive."

"Well, it was good advice to begin with," said Tom, "but

enough of this love-fest. I'm not sure the team will keep this up despite the fact that they could easily do much more. What else have you got for me?"

Thinking about it, Tim replied, "Let's think about the value variable. What can you give them that they value? How can you reward them?"

"Do you mean pay them more?" asked Tom.

"Can you do that?"

"Well, no," admitted Tom. "Not unless I want to drain my own bank account."

"Then don't bring it up," said Tim, "but no worries. Money does talk, but it's not the only speaker in this conversation. There's something out there that most people value more than cash—recognition. Simply be aware when they do something right and recognize it in a timely manner—not next month or next week but that day. A person's pride can feed off a sincere 'awesome' or 'job well done' for a long time, while a cup emblazoned with the company logo or even a certified check doesn't provide the same bang for your buck."

"That's an awesome point, Tim."

"Thanks," Tim said warmly, oblivious to the immediate use of the strategy he had just recommended.

"I really like it," said Tom. "If it gets me out of my own office a little more, that's great. I like one-on-one conversations more than those weekly meetings anyway."

"You're lucky. Many managers are promoted solely on technical skills and find the interpersonal part of the job difficult. Since you are so good at it, start using the Games and Goals strategy too. You know the story of the bricklayer, right?"

"Umm, remind me," said Tom, not willing to admit he hadn't heard of it at all.

"It's short. When two bricklayers were asked what they were doing, the first bricklayer replied, 'Building a wall.' The second took his time and, after some thoughtful reflection, responded, 'Building a cathedral.' You want to instill the bigger picture, why what they do matters, because if you do . . ."

" . . . all my dreams will come true," said Tom. "I see what you're doing—giving me the bigger picture. Got it. Timely recognition and frame the picture—communicate why what they do is important."

Tom allocated an hour a day to walk around the office, checking in on people to see how they were doing. If they impressed him, he told them so and sometimes did a little bit more; when one of his employees did a brilliant job presenting, he spontaneously offered to buy her lunch that day. Explaining the significance of the work was a little bit harder. He found that his employees needed the bigger picture to be framed in different ways. For some, how it would help their career made sense; for others, it was positioning an assignment as a symbol of responsibility; and for still others, it was about how their work affected their colleagues. Finding the right frame for the right person was a bit of a puzzle, but he got it correct more often than not. With one difficult employee, he explained it this way: "When you are done with your piece, it passes on to Suzanne. If you are late, she has to stay here late, which means scrabbling to find someone to pick up her kids from daycare, feed them, and put them into bed. You finish early, you make Suzanne's life easy. You finish late, you make her life hell." He never had trouble with that employee again. For good measure, he also tried to respect his employees' chronobiology and energy levels by instituting some *flextime*. Looking up some research, he found that just as students improved by a letter grade when allowed to sleep in by an hour, corporations that enabled flextime, allowing their employees to show up later but stay later in return, saw a nice bump in work performance.[3]

One night after work, Tim picked up Tom for dinner at a favorite restaurant. After they were seated and had ordered their food, Tim asked, "How's the leadership thing going?"

"Great," bragged Tom. "This transformer leadership is a snap."

"That's transformational leadership," said Tim. "Transformer is a type of robot, like Megatron or Optimus Prime."

Tom had been joking, but he corrected himself, "That's right, transformational and transactional." Then, he quickly changed the topic. "Speaking of which, you never told me about transactional leadership."

"Well, most people tend to favor one style or the other," said Tim, "but the best leaders have a combination of both. Transactional leaders excel at making plans, assigning tasks, *and* goal setting."

"Aah! This makes so much sense," said Tom. "I never knew what a pain my procrastination was to other people until I had to deal with procrastinators myself. Goal setting worked for me and it'll work for them too."

"Yup. That's what transactional leaders do. They divide distant deadlines into a series of short-term, specific, and realistic goals for their employees. Of course, too many goals and you become a micro-manager, also known as a control freak."

"I'm in no danger of becoming that. But still, how many goals do I need to set?" asked Tom.

"There's no firm answer on that," admitted Tim. "Essentially, people work hardest as the clock runs out, so you want to set as many deadlines as practical. At least have regular meetings, where you review people's progress and set new milestones. Keep in mind that some are already self-motivated and don't need much, while others need a lot."

"Yeah, I'm thinking of a few people who could benefit from minute-to-minute goals," said Tom.

"Just don't do what I have seen your company do," said Tom, as he began mimicking a pompous corporate voice: "'We want to raise revenue by 20% by year end!' That never works. I don't even know why they bother."

"I know what you mean. It's so distant and abstract, nobody can get motivated by it. Also, I don't know if anyone thinks it's even realistically achievable, especially with the downturn in the economy."

While looking over Tom's shoulder to see if their food would be arriving, Tim said, "Last year, your company made the goal too easy. When it's too easy, people do what they do when they cross any finish line—they coast, leaving extra performance on the table."

"Just like my initial goal," said Tom, "where everyone beat my schedule by *exactly* one day. I thought it was suspicious. I guess it's time for me to set the bar a little higher."

"When you do—and if I know you, you are going to love this—try partying," said Tim.

"You've got my attention. Go on."

"Don't forget to have a party at the end when you do accomplish it," said Tim. "People remember two things about a task: its best moment and the end moment. A party at the end will make it seem all worthwhile."

"I get it. Like good food at the end of a long conversation," said Tom, noticing that the waitress had finally arrived with their order.

Tom started to incorporate effective goal setting. When asking what his employees were up to, he kept pushing them to make concrete, short-term, and challenging goals. When he met them later, he had them give him updates on their progress. Some were naturals at this and used the opportunity to brag, which was fine with Tom—he was paying them with recognition. Others needed to be coaxed. Finally, he set a big group goal: they were going to cut production time by a week this month and if they could do it, which he told them he was sure they could, they were going to cut out early the following Friday for a party. Babysitting for parents and cab rides home for everyone would be on the company's tab. For the rest of the month, his team worked with a purpose and met the goal. The party was fantastic—as much a reward for Tom as for his team. He loved parties. In fact, whenever "his crew," as he started to call them, looked as if they weren't going to meet a goal, he doubled his efforts to make sure they met it and won the blow out. "Next time," he thought, "I'll put some money in the budget for a white-water rafting extravaganza. I can probably expense it as a team-building exercise anyway. And a prize for whoever gets the most reports out this month too."

Just when Tom was starting to get comfortable in this role as leader and manager, word came down from higher up. Unlike most other department heads, Tom was getting his budgets in on target and did his performance appraisals ahead of time. His performance was exceptional and his department was consistently the most satisfied and the most productive in his workgroup.[4] Inevitably, he was to be promoted once again. The secret to Tom's success was simply learning that what motivated other people was pretty much the same as what motivated him. To follow in his footsteps and become a better leader, you need to do the same. Good leadership is a skill that the world eagerly, even desperately, wants you to possess.

A WORD OF WARNING

Eddie, Valerie, and Tom benefited from enacting the principles of the Procrastination Equation, repeatedly hitting the three key components of expectancy, value, and time. When you put into practice the suggestions put forth here, you will benefit too. Just don't overdo it. While procrastination can lead to an inauthentic life, in which long-term dreams sour inside you, so can our efforts to completely eliminate procrastination.[5] A genuine and autonomous individual seeks a life endorsed by the whole self, not just a fragment of it. Trying to squelch your impulsive side entirely is ultimately self-defeating; the wants and appetites that propel a life depend upon being attended to. Overregulation—seeking the perfect over the real—isn't healthy and won't make you happy.[6] You are going to have to find a balance.

Just as the Procrastination Equation's techniques can work too well, so could the techniques in Will Ferguson's fictional self-help book. In his novel, after people read *What I Learned on the Mountain,* they did become blissful, contented, kind, and vice-free. They replaced their cigarettes and alcohol addictions with hugs and self-acceptance and swapped their oversized cheeseburgers with sensibly sized ones made from tofu. But all this virtue came at a cost: though everyone was equally content, they were also equally bland, interchangeable, and forgettable. Their personalities were whitewashed by their yearning to overcome all their flaws, and along with their vices so went desserts, fashion sense, and desire.

Procrastination represents a single swing of the pendulum, an emotional short-sightedness that sees only the present. As the pendulum swings to the other side, rational far-sightedness can become equally troublesome; we tend to focus only on the future.[7] When asked about their past regrets, workaholic employees wished they had occasionally goofed off, and exceptionally industrious students regretted studying through Spring Break.* Consequently, optimal self-control involves not the denial of emotions but a respect for them.[8] Not all indulgent delays are irrational. You need to have moments of expression, when you can laugh freely with friends, or let yourself go to be indulged and pampered. Using the words of W. H. Davies, a vagabond Welsh poet of my mother's youth: "What is this life if, full of care, we have no time to stand and stare." To be idle, frivolous, spontaneous, and whimsical—these qualities deserve a place in our lives too.

LOOKING FORWARD

Nine thousand years ago, procrastination didn't exist. Back then, if we worked when motivated, slept when sleepy, and acted on other urges as they came upon us, we did so more or less adaptively. In that golden age, our compulsions fit our daily demands like jigsaw puzzle pieces. We were designed for that world, life before the invention of agriculture. Fast forward nine thousand years and that same human nature has equipped us with inclinations that are ill-suited to the everyday. We have to-do lists filled with diets, early wake-ups, and exercise schedules, among a host of other ugly and motivationally indigestible ordeals. Almost every aspect of our lives reflects this maddening mismatch between our desires and our responsibilities, as we overemphasize the present and sacrifice the future. We overindulge in the immediate pleasures of fats,

* As psychologists Walter Mischel and Ozlem Ayduk wrote: "An excess of will can certainly be as self-defeating as its absence. Postponing gratification can be an unwise and even stifling joyless choice, but unless people develop the competencies to sustain delay and continue to exercise their will when they want and need to do so, the choice itself is lost."

sugars, and television, putting off dieting and exercise. We let loose anger and rage, putting off needed reflection and reconciliation. We have predilections toward the easy pleasure of promiscuity, risking long-term relationships and reproductive health for the forbidden but immediate. Each of these examples reflects a nature that was once adaptive but is no longer, a nature that outrageously values the now more than the later. The story, however, doesn't need to end here.

As *The Procrastination Equation* stresses, irrationally putting off is a tendency, not an inevitability. If we can accept our internal state of affairs, we can counter it. Instead of believing we have the temperament of divine beings, we can reconcile ourselves with our humanity—to the fact that we are flawed and compromised creatures—and act accordingly. We can make procrastination an irrational delay of yesterday, what we all once did or didn't do, but only if we acknowledge our own limitations and adopt advice consistent with this understanding. To put it all into practice, you don't need to ask permission. There will be no handwritten invitation. To live your life as you always wanted to, to be the person you always wanted to be, you know what to do. You are holding all the answers in your hands. Now do it.

POSTSCRIPT

Procrastination's Chapter 11

The beauty of procrastination is its ubiquity; tracking its scent leads into dozens of scientific fields. If you duplicate my path, you will start with psychology, where the bulk of the work has been done, but you will quickly find yourself in economics, which is becoming a dominant force on the topic. You will take a stroll through the applied issues, like retirement or debt procrastination, perhaps taking a peek into the legal implications, such as suggested bankruptcy laws. From economics, you would naturally wander into neuroeconomics and become interested in the neurobiology of procrastination, a detour that, of course, would give you a chance to look at the basis of all biological study, evolution. You would learn that procrastination is a common and consistent human trait, one we share with species across the animal kingdom. Then, instead of considering where we came from, you might reverse your perspective and see where we are going, getting into societal issues, especially long-term concerns like environmental degradation. If you start wondering why the government doesn't do more, you will discover that they and other organizations have procrastination problems of their own.

Having been studied in so many disciplines, procrastination has become a Rosetta Stone, where the same phenomenon is translated into a dozen tongues. This pool of resources allows us not only to translate findings from different fields, such as from economics into psychology, but also to form a common language of human behavior, an Esperanto of the social sciences. It's an important accomplishment.[1] As Christopher Green concludes, writing in psychology's premier journal, "it [integration] would doubtless be considered the greatest scientific victory in the history of the discipline," one that can rescue psychology from the realms of a "would-be science."[2] And if you can integrate psychology with economics, sociology, and biology too, even better. This was actually my original purpose in creating the Procrastination Equation—to help integrate the social sciences.[3]

Unfortunately for procrastination, its pervasiveness makes it an obvious target. Having a common basic model, one that each discipline can adopt and customize, could be incredibly dangerous to our familiar enemy. Integration enables exponentially more progress in all disciplines. This understanding has already permitted the physical sciences to provide an endless stream of game-changing advances, from the laptop I am using to write this book to the nuclear energy that powers the electrical lines.[4] By working from a common model of reality, the physical sciences share and pass knowledge across specialties and research foci. Similarly, such synergy could supercharge the social sciences. Herbert Gintis, an economics professor emeritus from the University of Massachusetts who has long argued for integration, concludes: "The true power of each discipline's contribution to knowledge will only appear when suitably qualified and deepened by the contribution of the others."[5] You see, it is all connected, all of it, as we are all studying the same thing: people's decision making and behavior.[6] As one area informs the other, our fight against procrastination necessarily gives insights into reducing obesity, building better retirement savings programs, and much, much more.

Once disciplinary integration comes about, we will have gone a long way toward truly mastering our own minds. As it currently stands, we as a society can do better. Consider that the top two ways that people procrastinate are through their televisions and through their computers—about a quarter of their waking hours in some parts of the world. People seeking help to curb their addictions freely acknowledge that they use these temptations to excess.[7] Because TV watching has been associated with the rise of obesity and the erosion of the family, huge efforts have been put forth to reduce our consumption.[8] Nothing has been truly effective; the hours used and incidence of abuse rise yearly. If we adopt a more integrated viewpoint, using some of the principles from *The Procrastination Equation*, we can change this. We only need to apply the principles of self-control to our own technology.[9]

When I watch too much TV, I blame my digital video recorder (DVR). It makes it easy for me to find a show I like and watch it when I want. Naturally, the easier it is to find good programming and the faster it can be accessed, the more I will make use of it. You will too. Though DVRs are part of the problem, they are going to be part of the solution too, as they are also the perfect platform to enable self-control techniques. Self-control improves when we receive accurate feedback about our behavior, which we can then use as reminding cues and to help us set goals (see *Scoring Goals* and *Making Paying Attention Pay*). An add-on for a DVR could be a prominent digital display that reflects how much TV you have watched today or this week. As you see the hours visibly rise while you watch, so will the desire to turn off the set. The DVR could even track your long-term viewing, calculating when and what you are watching.

Also, DVRs could permit precommitment. Devices are available to enable parents to limit the viewing habits of their kids, but there are few options for parents themselves. With a DVR, a series of precommitment measures could be incorporated. The first few could just be devices for enabling delays. A long code, for example, could be laboriously inputted before viewing. Alternatively, it could lock you out for a few minutes, or perhaps require confirmation multiple times, giving you a chance to have second thoughts. As delay lengthens and impulsive choices become impossible, you should be able to make more rational use of your viewing time. If this isn't enough, you could lock yourself out temporarily, perhaps only allowing viewing within given time periods or up to a total number of hours each day. Best of all, whichever of these options we as viewers want to activate, if any, the choice—the intention—is ultimately ours.

For Internet procrastination, similar solutions are already on the market. Attentional control programs like *RescueTime,* which let you see exactly what you have been doing with your day, are freely available. As an added feature, *RescueTime* also assists in goal setting and permits the creation of comparison work groups, thereby activating the *Vicarious Victory* principle. Cyberly seeing others hard at work should inspire, or at least spark, your competitive spirit. Furthermore, *RescueTime* allows you to voluntarily block your own access to the Internet for chosen periods of time, permitting a precommitment strategy that eliminates distractions. If this could be complemented with a sophisticated and difficult-to-subvert nannyware program—like *Chronager,* except self-administered—it is hard to imagine a more effective self-control platform. Right now, the pieces are all there; we just have to bundle them together.

These tools for rationalizing television and computer use could be easy to build and implement. Though not yet fully developed, they have almost coalesced. When they are finally built, the market is virtually everyone, but definitely the chronically procrastinating quarter of the population. These tools would have society-wide effects and an observable impact on national GDP; if they cut procrastination even by half, that would amount to trillions of extra productivity each year worldwide. With further advances in integration, more such tools that address our own weak wills should become commonplace, designed into our society's fabric. And ironically, for all this, we can partly thank procrastination. Fittingly for an irrational self-defeating delay, by making possible the groundwork for integration, procrastination may have contributed to its own defeat.

WORKBOOK

A Step-by-Step Guide to Getting Stuff Done

The book you hold in your hands is our best present understanding of motivation, an enigmatic science that thousands of researchers have attempted to unravel for centuries. Some of the insights here have become staples of the self-help field, while others are still relatively unknown, but if I have done justice to the science of those insights, I hope I've left you thinking, "Great! I'm convinced." If not, I'm ready to give it one more very practical crack, because, after all, the key to successfully motivating yourself will be in how you *implement* the techniques I've described.

This can be tricky. Making the leap from abstract idea to concrete application isn't easy for everybody in every situation—especially for procrastinators. Sometimes you need a little more inspiration or practice, just as you'd practice a new dish from a cookbook, for example—your first attempt may look nothing like the one in the photo of the recipe. That's why you keep at it. (But we'll try to keep your motivational meals from being burnt or over-spiced the first time around.)

Practical application is also difficult because the way you procrastinate and what you put off are personal: like proverbial snowflakes, no two people have the exact same pattern in the tasks they put off. And if our lives contain a thousand tasks, then there are a thousand ways to procrastinate. Maybe you're putting off asking for a job promotion, for example. Or perhaps delay is costing you in your home life: are you getting on with your renovations or repairs, or just putting a bucket under the drips from a leaky roof? And what about your relationships? Are you investing in your friendships and carving out time for romance, or just penciling it in for tomorrow? When we multiply all the ways we procrastinate with the many ways each technique can be used, we enter a snowstorm of permutations.

Still, a few good examples can go a very long way. If I can't prepare you for every situation, I can at least walk you through a few common ones. I've picked three widespread issues that plague procrastinators: dieting, working (with a focus on writing), and saving money. For each, I will provide specific solutions, translating the principles of motivation into real tasks you can do today. Let's get stuff done.

DIETING

How is your diet going? Chances are you are on one or that there is a diet in your future. The majority of Americans think about dieting all year round, with as many as 41 percent on a diet at any given time, in an attempt to lose an average of thirty-seven pounds.[1] Britons aren't far behind: about a third are constantly on a diet. The statistics for the rest of Europe, even France, are similar.[2] Most telling, increases in obesity and dieting are appearing in Eastern Europe, India, and China as they move to the free-market system.[3] The International Obesity Task Force notes that obesity rates have typically doubled every decade, and for many European countries, they have tripled since the 1980s.[4] There is no way to sugarcoat this: pandemic obesity and massive rates of dieting mean that we as a civilization are failing at self-control.[5] With tomorrow being the best day to lose weight, odds are we will let ourselves become heavier despite our intentions to become thinner.[6]

Trying to lose a few pounds is clearly a good thing when you're overweight. Obesity drastically shortens lifespan, leads to a host of disabling health problems, and can be socially stigmatizing. But despite this litany of well-known ills, willpower often fails us and we stumble. In the food fight between the long-term goals of our prefrontal cortex and the here-and-now temptations of our limbic system, the game is rigged for the latter. Let's tip the scales in the other direction; dieting is about to become much easier. Here's a walkthrough of how to apply the Procrastination Equation to take off or keep off the pounds.

EXPECTANCY

1. Success Spirals

- Aim to lose at most two pounds a week, tapering it down to one pound a week as you approach your goal weight. This is realistic, and by not overreaching you are setting yourself up for success. If you make super-sized goals, you'll feel discouraged when you don't reach them.
- Pay attention to intermediate or process goals as well, not just the outcome goal of weight loss. If you improve your nutrition or exercise habits, that counts even if it doesn't immediately show on the scales. There is a huge health difference between those who exercise and eat well and those who don't, even if they weigh exactly the same.[7]

2. Vicarious Victory

- Start watching TV shows like *Heavy* or *The Biggest Loser*, both of which have inspired millions. And start reading social networking sites or blogs like Shauna Reid's *The Amazing Adventures of Dietgirl* (www.dietgirl.org) or Alan Spicciati's the *Pounds Off Playoff* (www.poundsoffplayoff.com). Online, there are armies of people wanting to connect and share their successes and difficulties. Seeing others succeed helps to spark belief in yourself. And the results are demonstable: eating better and exercising regularly have predictable effects.
- Spend more time with friends who have lost weight or are keeping an active, healthy lifestyle. Good habits are as infectious as bad ones.

3. Wish Fulfillment

- Imagine that feeling of vigor that will infuse your body and all the activities you'll engage in with friends and family once you're in shape. As a parent, for example, it might be playing with your kids at the park again. Now contrast that image with where you are now. Are you tired and flabby, spending far too much time in front of the TV? If so, it probably doesn't feel good, does it? But it does make you want to do something about it.

4. Plan for the Worst, Hope for the Best

- Anticipate that life doesn't proceed perfectly, and have an emergency plan in place for when the wheels fall off. For example, is there a personal trainer to book or a friend to call and get you back on track if you fall off your diet for four days?

5. Accept That You Are Addicted to Delay

- Be aware of your danger zones. Do you find yourself overeating in the same places or situations, like buffets? If so, try to avoid them entirely.

VALUE

1. Games and Goals

- Write down and affirm all the good reasons to lose weight, using the twelve major life domains from the table in chapter five for inspiration. Is it being a more attractive romantic partner? Perhaps a better parent? How about improving your feelings of self-worth or living long enough to enjoy retirement? Connect dieting to your life's big picture.
- Make yourself publicly accountable by advertising your intentions. Just make sure your goals are specific and verifiable, such as number of pounds lost. For example, consider the Withings WiFi Body Scale. Every time you step on it, the scale tweets your weight. The opposite strategy—conveying vague intentions like "I'll eat better"—can be detrimental to the cause. You will get unearned satisfaction from making aspirational but unverifiable statements, which will in turn sap the motivation that would have driven you to actually follow through.
- Focus on the approach. This is about pursuing healthier options and eating proactively. The focus is on what you do rather than what you don't do. It is about eating a greater proportion of vegetables, not just cutting calories and carbohydrates. How many salads can you have for dinner this week?

2. Energy Crisis

- Snack between meals, but don't get too excited: snacks are only as healthy and as high-fiber as you can make them. Don't allow your blood glucose to get too low—you will start making bad food choices.
- Get to bed at a regular and earlier hour. Food often takes the place of sleep as you try to replace rest with calories.

3. You Should See the Task I'm Avoiding

- If you give in to a food temptation, minimize the damage. Have half a cookie rather than a whole one if you feel you must give in. In other words, allow a small temptation if it prevents a larger one—less is always more.

4. Double or Nothing

- Go on diets together with other people. Dieting is more enjoyable that way, and if you have a partner to share your efforts and struggles with, you are much more likely to keep at it. Besides, a little competition helps spice up any activity.
- Buy an outfit now that will fit you in six months as a long-term reward for staying on track with your weight loss.

5. Let Your Passion Be Your Vocation

- Figure out which healthy foods you like best. Some are delicious and some aren't, depending on your tastes. Just follow the lead of the British celebrity cook Jamie Oliver. He has made it a life mission to show that nutritious meals can be tasty and cost-effective. You can do the same for your friends or family.

IMPULSIVENESS AND DELAY

1. Precommitment and Self-Binding

- Create some sizeable penalty for failing to meet your weight-loss goal. Get a "referee" who can verify whether you meet your goals. If you fail, the referee ensures you pay the penalty, which can be as harsh as you want. Donating a week's salary to an anti-charity, a cause you detest, will more than suffice. See www.stickk.com for more pointers.
- Plan for your next meal well before you eat it, when you are already full. Know exactly what you are going to prepare. If you want to indulge in a treat, make that decision at the previous

meal, not in the moment and when you are hungry; that's when the limbic system rules, and it thinks diets are dumb.

- Precommit by sating your appetite with a garden salad at lunch or dinner. Give it a twenty-minute chance to assuage your hunger before you think about the main course. You will find your eyes will stop being bigger than your stomach.
- Or, precommit when dining out by telling the waiter from the outset not to tempt you with a dessert menu at the end—and make the tip contingent on honoring this request.
- Employ the time-honored technique of food destruction. The fitness guru Jillian Michaels pours candle wax on her food when she has had enough, but you can use a dollop of salt instead.
- Take any unexpected treats to your office or school and share. People will love you for it.

2. Making Paying Attention Pay

- Make healthy snacks the easier and more visible of your food options. If you have healthy snacks by your side, you will choose them because they are the most accessible.
- Place tempting alternatives like the candy bowl or the doughnut box far away so you have to step away from your office or at least open the fridge door. If you can't put them far enough away, hide them. Tupperware is your friend.
- Write your grocery list beforehand to avoid regrettable impulse purchases of pre-packaged and easily consumed treats. Even better, have a grocery service deliver your food for you.
- Mentally deconstruct food into its basic components, even if you don't record the calories. Many dieters don't commit to counting calories, but seeing food in abstract terms helps diminish the power of temptations. That way, every temptation is merely so many grams of sculpted fat and tablespoons of flavored sugar.

- Take a moment to reflect on how you will feel later about giving in to food temptations. Remember that most of the time the food really doesn't taste as good as you expected—and it isn't worth the calories.
- Tape a picture of yourself at your heaviest on the fridge.
- Use covert sensitization by reflecting on what horrible maltreatment could occur to your food behind the scenes.
- Control your portions. The size of your plate cues you how much to eat. You will want to fill up your dish, so instead of putting your salad in a bowl, place it on the same plate as your main course. If that's too messy, keep the salad bowl but put your main course on a smaller plate.
- Serve yourself in the kitchen and then eat elsewhere, so that you physically have to go back for seconds. Getting up to refill your plate creates a barrier that is usually big enough to decrease your consumption.

3. Scoring Goals

- Have a weekly weigh-in first thing every Saturday morning to see how you are doing.
- Add pieces to your weight-loss regimen incrementally. It makes the diet less daunting. For example, make a goal of cutting out snacking this month, desserts the next. Eventually, as you add more strategies, the weight loss will come.
- Make sure that when you plan your meals and snacks ahead of time you do it in detail. You need to specify not only what you are going to eat but how much and when meals or snacks start. If you are going to a restaurant, for example, check out the menu online to predetermine your meal.
- Repeat after me: "When I get a food craving, I will eat a healthy snack." Use similar implementation intentions as needed.

- Make when and what you eat predictable. The aim is to foster a routine so that eventually it happens automatically and you won't have to consciously rely on all these techniques. You will be using them reflexively, with all the effort it takes to brush your teeth or tie your shoes.

WORKING

For many of us with bills to pay or degrees to earn, working invariably requires time in front of a computer. Computers are wonderful, multifunctional devices, but because they can do so much, they often allow us to do nothing at all. The journalist Nicholas Carr describes these Internet-enabled devices as "distraction machines," because it would be difficult to design a tool that makes it harder for us to concentrate on our work.[8] Essentially, in seconds we can transform our workplace into a virtual strip club, casino, game room, or movie theatre. Ever tried writing in a strip club? It is extraordinarily difficult.* Now that we have motivationally polluted our desks and offices, is it surprising that we have trouble getting anything done?

Most people report procrastinating at least a quarter of their working day away.[9] When you add up all that dilly-dallying, you get three months of wasted time annually. It's the equivalent of waiting every year until April Fools' Day to start working, except the joke is on us because it's our performance getting pranked. The more we procrastinate, the lower our pay and the fewer our promotions.[10]

Let's put some productivity back into our workday. Though a variety of vocational activities can take place in front of a computer screen, I will focus on writing, which seems to bring out the procrastinator in us more than any other pursuit.

* Or so I imagine.

EXPECTANCY

1. Success Spirals

- Build up your stamina. How long can you generate concentrated work while at the screen? Aiming never to procrastinate or goof off while at work is a praiseworthy goal but rarely achievable, so as long as you are improving your work endurance or your ability to focus, you should be proud.
- As always, pay attention to intermediate or process goals as well, not just the final, long-term goal. Professional writers often attend to both hours of work and words written by saying something like, "I'll stop after a thousand words or five hours, whichever comes first." This approach rewards you for having a particularly productive day, as you can skip out a little early, but makes sure that you still work even when the words don't flow.

2. Vicarious Victory

- Get more involved in your professional association; we gain inspiration and confidence by rubbing shoulders with optimistic others. You will find those who volunteer are often encouraging. Besides, making connections could be great for your career as well.
- Join a mastermind organization that focuses on personal and professional development. And note that this is a common strategy for those already brimming with success: many CEOs and entrepreneurs belong to organizations like TEC (The Executive Committee) or EO (Entrepreneur Organization). If you are not part of the C-Suite, look for your local writer's group or Toastmasters club.
- Create a healthy and supportive circle of coworkers. This is easy—and free. A monthly lunch where people share their successes and provide advice and support is often enough.

3. Wish Fulfillment

- Mentally capture that feeling of unfettered leisure time you had when you were young—when you had nothing else to do but replenish and enjoy, when your play was unencumbered, and when there was no thought of what else you should be doing. Remember when a summer's day, a pile of sand, and a garden hose were enough? Now contrast that feeling with where you are now. By procrastinating, your downtime is darkened by the cloud of unfinished activities, spoiling whatever pleasure you should be fully immersed in. Doesn't feel good, does it? So do something about it.

4. Plan for the Worst, Hope for the Best

- Remember the planning fallacy? Everything takes more time than you expect. This is one of the hardest biases to overcome, so don't expect any of your "time until completion" estimates to be accurate. Routinely, hunts for files or facts that would take "just a second" are still underway an hour later. Estimating how long tasks truly take is an art, with bigger projects needing bigger corrections. So plan for the worst and start doubling or even tripling your time estimates. You can hope it will take less time, but it probably won't.

5. Accept That You Are Addicted to Delay

- Perhaps you aren't one of those people who can surf the Internet and keep it in perspective. If you can realistically assess your limitations, it allows you to act effectively within them. When choosing among all the self-control techniques outlined in this book, choose ones appropriate to your level of need. If you are a heavy-duty procrastinator, consider a heavy-duty technique like precommitment. There is no shame here, because you aren't the only one. For example, the Yale Law professor Ian Ayres has his school administrator install nanny software, specifically Cybersitter, to prevent his web surfing. The science fiction writer Robert Sawyer has installed Internet Access Controller—and given the password to his wife.

VALUE

1. Games and Goals

- Answer this: why are you in front of the computer in the first place? What is this really all about? Answer correctly and you have reconnected with your core motivations. The value of work isn't entirely inherent in the task; you can influence it by how you do it and how you see it. Framing your work in terms of advancing your meaningful life goals means you will experience some of the luster from those goals in the here and now. Working efficiently will allow you to make gains across all major life domains, even if that's just being able to leave work earlier and guilt free. Write down reasons for getting stuff done that resonate—clearing time for friends, for instance, or mastering your craft.
- Tell everyone who matters when they can expect your work to arrive. By making yourself publicly accountable, you raise the stakes on the failure to follow through. Just make sure to be specific in your deliverables, because you are likely to leave undone everything unspecified or implied. Promising to get to it later isn't really promising anything at all.
- Focus on increasing your hours of productivity instead of dwelling on all those wasted hours. This is the difference between approach and avoidance, with the former being the more motivationally powerful. How many hours of productive work can you eke out of yourself today or this week? Get closer to your finish line, not farther from the starting line. Don't discount small gains, as over the course of a year, a 10-percent

improvement in efficiency translates into more than an extra month of getting stuff done.

- Consider getting time-management software. My favorite is RescueTime, and there are a few features I particularly like about it. The main one is that it lets you keep score and compare your productivity to everyone else. This provides the basis for a competitive game that you can play against yourself or your friends. Play for money or just bragging rights. This will work. If Tom Sawyer could get his friends to pay him to paint his aunt's picket fence, there's a way to up the value of almost anything.

2. Energy Crisis

- Fuel yourself to fuel your work. Willpower is closely linked to your blood glucose level. If sugary treats are your snack of choice, be prepared to eat a lot of them. The dietary equivalent to burning gasoline, they release all their energy in an instant, requiring you to refuel moments later. Try some complex carbohydrates instead; like a difficult math puzzle, your body takes a little longer to (dis)solve them, so they provide a slower and longer release of energy.
- Get plenty of sleep. Being too tired is the number one reason people cite for putting things off. If you are not getting eight hours of a sleep a night, you are harming your performance during the day. Lack of sleep significantly lowers all forms of cognitive performance, making everything you do more difficult. Here's an idea: instead of setting your alarm clock to get yourself up, set it to cue you to go to sleep. When you hear that buzz, start your wind-down routine.
- Make better use of the energy you already have. Better sleep and regular exercise are just a few ways to increase your energy level, but too few people optimize the energy they've already

built up. Allocate your worst tasks to when you have the most energy. Your peak productive hours will shift depending on your chronobiology, but most people's power hours are between 10:00 a.m. and 2:00 p.m. Dedicate these four golden hours to tackling your most difficult work. If you can truly dedicate this time to real work, you can often coast for the rest of the day without anyone noticing.

You Should See the Task I'm Avoiding

- Try "productive procrastination" instead of playing *Minesweeper* or *Angry Birds*—the kinds of activities that only maximize the cost of procrastination because you really are doing nothing of value. If you can't bring yourself to work on a task, use it as an opportunity to work on some other activity you have been avoiding. When you compare this task to what you truly dread, the second-best seems like a pleasant option. With something you hate looming over your head, this is often the easiest time to make headway on everything else.

4. Double or Nothing

- Reward yourself after completing a task, or pair the task with a reward, such as indulging in a specialty coffee. This can make what was once difficult bearable. Just make sure the reward doesn't take the place of the work itself. For example, there is very little you can effectively do while simultaneously watching YouTube. A little music, on the other hand, probably won't get in your way.

5. Let Your Passion Be Your Vocation

- Ask yourself: if you just can't feel motivated by your work, might the problem be the work itself? We put off what we hate or find boring, and do what we love. Letting your passion be your vocation is the most straightforward technique to translate into the workplace, so let's consider what we can do.

i) First, "don't retrain, reapply." A bad supervisor can make an otherwise suitable vocational choice seem awful. So if you truly feel this is a career and not just a stop on your journey, try to get a better boss in the same career field—it makes for a better job.

ii) Second, recognize whether it's the work or the career that's in your way. The early stages of most careers aren't glamorous but grunt work. However, if we stick with it, job satisfaction tends to increase with advancement. This is because every job has a host of tasks, and often we are only well suited to a subset of them. Do you really like doing timesheets? As we advance, we often gain the autonomy to angle our career toward the aspects that we enjoy, allowing us to better define our jobs. And once you have made it to a managerial level, explore the benefits of delegation. Getting good at it is one of the enjoyable aspects of leadership.

iii) Finally, determine if your procrastination may indeed be due to a poor career choice. Do you simply hate or are at least disinterested in what you do? No shame here, but if you have the opportunity, get it right next time. Not everyone has the capacity to switch career tracks, especially if the economy is bad and their level of debt is even worse. To the extent that you are able, take a little more time finding a career that you can be passionate about. Doing work you care about is what makes for a good life. If you are considering your college major, a little extra advice from companies like Career Vision is money well spent. There are many thousands of people in poorly chosen careers that sap their motivation from the moment they start their working day.

IMPULSIVENESS AND DELAY

1. *Precommitment and Self-Binding*

- Since most distractions are online in origin, stem the problem at its source: hinder or disable your Internet connection. It's not always easy to act in your best interests when you are already in the grip of temptation. Writer Jonathan Franzen permanently closed his laptop's Internet cable connection with Krazy Glue. His colleague Jonathan Lethem goes one step further: "I've set up a second computer, devoid of Internet, for my fiction-writing. That's to say, I took an expensive Mac and turned it back into a typewriter."
- If you're unwilling to mutilate your laptop like Franzen, temporarily disable the Internet through the use of some self-control software. You have a lot of choices depending on your operating system. For example, programs like Internet Access Controller and Self-Control allow PC and Mac users, respectively, to temporarily shut down both their e-mail and Internet access.
- Erase all your games from your computer, along with any quick-launch icons.
- Make the passwords to all your social networking sites somewhere between thirty and fifty digits long. Don't have the computer automatically log you in; you have to input this gargantuan password manually each time. With such an upfront cost, you'll surf only when your desire to take a break crosses a meaningfully large threshold.
- Commit to playing more outside of work (the most fun of all these techniques). If you don't meet your needs outside the office, you'll feel the urge to do so at work, often inappropriately.

2. *Making Paying Attention Pay*

- Avoid multi-tasking, especially with e-mail. Turning off the ding and other indicators for your e-mail is easy and effective.

- Separate your leisure apps from your work apps. Ideally, you would do all your non-work activities on a separate computer. Failing that, the Mac has a wonderful program, Spaces, that lets you create up to sixteen virtual desktops, each with their own look and bundle of apps. PC users can do the same by simply having a separate log-on profile. When you want to relax, log off of your work persona and on to your play persona, which should have a distinct look and feel. You can even add a little precommitment here by adding a long or cumbersome password to your play persona.
- Put distractions at a distance to discourage unwanted interruptions. Leave your office door open only a crack and post a warning sign (e.g., DEEP IN THOUGHT) outside. Angle your chair so people can't make easy eye contact, the first step toward conversation, as they pass by your office.
- Try some covert sensitization by imagining all the malware that is likely being installed on your computer because of your Internet surfing, especially if you are part of the nine-to-five crowd that accounts for 70 percent of adult website traffic. Try explaining *that* to your IT department.
- Add reminders that help you focus on your work. These work cues can be anything, from to-do lists to a picture of your hero or role model. Screen savers set to activate after a short hiatus, say five minutes, work particularly well, as they can flash your favorite motivational quote.

3. Scoring Goals

- Break plans down into tasks. Most people operate well with daily or weekly goals, especially specific goals. For tasks you are procrastinating on, don't just plan what you have to do but also establish the exact time or situation that will cue you to start (e.g., after my coffee this Monday morning). Furthermore, fostering a regular schedule, with predictable starts and stops, allows any aspect of the work environment to become a cue in itself. If you want the sight of your desk or the strike of the clock to motivate you, try to start work at the same hour, the same place, and in the same manner as consistently as possible.

- Make implementation intentions to exploit small pieces of time. Life is filled to the brim with ten-minute chunks of unallocated time. Add them all together and stuff gets done.

SAVING MONEY

Some debt makes sense. It smoothes expenditures over the course of our lives, allowing us to own a home, receive an education, or go on a luxury vacation while we're still young. But when it goes too far, debt robs our future selves and steals the pleasures of tomorrow for the exclusive use of today.

Being procrastinators, we have, of course, gone overboard. All over the world, the amount of personal debt has been increasing for decades. In my home country of Canada, we owe about 150 percent of what we make (after taxes) on average, putting us in fourth place in the household debt stakes.[11] Switzerland, Britain, and the United States all have even higher levels of debt.[12] If interest rates remain low, we can sustain this, but that is a very big "if." Nothing lasts forever, good or bad, and the cost of borrowing money will invariably rise. When it does go up, so too do payments, financially pushing another chunk of the nation over the edge so they can no longer afford to service their debt.

If you want to sock money aside to avoid such a trap, you must develop a better sense about your money situation right now. You know that certain purchases need to be the exception rather than the rule and that all that credit card debt is killing you. I'm not here to give you basic financial lessons, but I do want you to be able to follow through on what you already know. Here's how.

EXPECTANCY

1. *Success Spirals*

- Try to save more. Easier said than done, right? But just start with one expenditure, such as food. Many families are able to lop $2,500 off their grocery bills every year by shopping a little more skillfully. They check the cost per unit, buy in bulk, comfortably ask for rain checks, and hit the discount grocers (which sell almost all of the *exact same* products as their upscale counterparts for substantially less). Once you get good at it, look to apply this same skill set to other major purchases in your life, from the clothes you wear to the vacations you take.
- After mastering how you spend your money, work on how you invest it. First, focus on consolidating debt and reducing the interest you pay. Only then should you consider if you can increase the return on your investments. Build your financial acumen one hurdle at a time. Don't worry about what you don't know about wealth creation. This is about gradually expanding what you *do* know about it. And as an almost absolute law of the universe, you will get better with practice.

2. *Vicarious Victory*

- Follow by example: the world has an abundance of smart shoppers and savvy everyday investors who want to welcome you into their fold. There are thousands of books and websites filled with suggestions on how to make a dollar go further and how to make it work for *you*. Of course, "buyer beware" still holds and you will find your share of scams, but stay safe by not straying too far from the basics. I suggest starting with books like *The Wealthy Barber* or *The Richest Man in Babylon*. In an encouraging and accessible format, they walk you through simple methods for reliably increasing wealth, using a few characters that most readers can relate to (i.e., their own Eddie, Valerie, and Tom).

- Consider joining an investment club, which come in all levels of sophistication. For example, Ellen's Degenerates Investment Club focuses on improving people's ability to properly handle their own money in a fun and inclusive manner. It caught my eye because it periodically reviews financially relevant books, including my own.
- Join bulk-buying clubs to rub shoulders with smart shoppers. They pool together their purchasing power and usually get good discounts because of it. Or, in a pinch, if you have one friend who knows where all the deals are, go shopping with him or her; it is like a micro buying club, letting you take immediate advantage of all those "2 for 1" deals. The corollary to this is avoiding your high-rolling purchase pals who assuage their own guilt for luxury purchases by coaxing you to buy big too.

3. *Wish Fulfillment*

- Mental contrasting or guided fantasy increases motivation, so let's apply it where financial procrastination is often worst: taxes. Think exactly about how you are going to spend your tax savings. What indulgence do you have in mind? Feels nice. Now contrast that thought with *not* having that treat, because you are not going to get it if you don't get going on your taxes.

4. *Plan for the Worst, Hope for the Best*

- Make your financial goals challenging but achievable. If you are overly aggressive in your plans to cut down debt or put cash aside, you will get frustrated when you don't hit your targets.

- Count on unpredictable emergencies and expenditures getting in the way of your financial plans. When preparing for battle, Dwight Eisenhower always found that "plans are useless but planning is essential." In other words, life rarely goes according to plan, but you need one anyway. So expect a few setbacks and focus on how quickly you can get back on track.

5. Accept That You're Addicted to Small Purchases

- Look at the big picture when it comes to your finances. We often fool ourselves into believing that our choices are isolated events, not part of an extended trend or larger pattern. In this way, we can whittle away our pay checks on small daily purchases such as specialty coffees, ATM fees, takeout dinners, or the nightly movie rental. Add these together and you get $25, more or less, which doesn't seem like much but adds up fast if you do it every day. Cutting these purchases out entirely could hand you more than $9,000 in extra spending money each year, and just cutting them in half will save you at least $4,500. Try multiplying each of your minor purchases by 365, or, for ease of calculation, just 300. Doesn't seem quite so small anymore, does it?
- Take another look at all your cable, Internet, phone, and other monthly utilities. This time, multiply the figure by twelve and ask if you really need it. Often people find they can do away with their landline and all those add-on cable packages. Alternatively, bundling these services together with one provider often gets you a discounted rate. In any case, you're likely to find that when you call to cancel or switch services, the company will give you some freebies in an effort to keep you as a customer.

VALUE

1. Games and Goals

- Consider putting money aside as giving a gift to your future self. A useful frame for saving money goes under the moniker "Pay yourself first."
- Avoid trying to keep up with the Joneses. Once you start down that road—and spend beyond your means to one up your neighbors—you will never have enough. "You're rich if you are making more than your brother-in-law," one joke goes. In other words, the value of wealth is relative.* Remember that today, any Average Joe or Jane lives in many ways better than the kings and queens of yesteryear.
- Be mindful of how you describe your attitudes about money. Do you see yourself as a penny-pinching tightwad or as penny-wise and economical? Alternatively, are you a high-rolling big spender or a squandering spendthrift? Remember that being good with money should be seen as a blessing. It's why providence means both "divine or godly guidance" and "saving for the future."
- Try turning your finances into a game. First, you need a way of keeping score, and this is where a monthly budget can be motivationally useful. Tracking how much you are bringing in and how much you are spending gives you the raw data. Then, with a predefined reward plan, you can allocate treats and splurges for yourself when you hit key milestones. In fact, many computer games have resource allocation or budgeting as a core part of the game—think FarmVille or almost any of the Sims games. It takes a little longer to get hooked in real life, as each turn takes place in real time with real stakes, but many users of personal financial software like Quicken find them addictive.

* I sincerely apologize for this pun, but I found it irresistible.

- Ask yourself if whatever else you are doing is worth more than $300 an hour. Completing your taxes takes about two hours with today's tax software and most people receive a $600 refund. On average, that's about how much you're paying yourself to do your taxes. If you owe money, you could save even more by avoiding penalties.
- Consider allocating any pay raises to your savings if you can't allocate any of your present income. This exploits the wage illusion, as a pay raise often just keeps us level with inflation. But it feels like more money and we find it easier to commit to saving it.

2. Energy Crisis

- Don't make big purchases when you are tired. Willpower is intrinsically linked to your energy levels. Sleeping on any major financial decisions should be the rule. For every last-minute offer you'll miss, you will avoid a dozen other regretful decisions.
- In fact, don't shop when you are tired or hungry, period. Fatigue and hunger leave you more vulnerable to sales pitches. Without internal stores of energy, you default to the sales staff's agenda, not your own.

3. You Should See the Task I'm Avoiding

- If you feel compelled to make a large luxury purchase, try making a smaller one instead. People do this naturally during hard economic times, in what economists call the "Lipstick Theory"; when times are tough, women will treat themselves to expensive lipsticks instead of expensive clothes. Science suggests that experiential goods in particular provide the best pampering. Go out to a ball game or lie down for a therapeutic massage at your local spa. It should be a little less pricey than a designer handbag or a high-tech set of golf clubs.

4. Double or Nothing

- Negotiate a spending and savings plan that you and your partner can both agree on, and you'll make sharing your finances much more pleasurable. The number one cause of divorce is money and how it is spent, so do not underestimate the power of appreciation! Mutual recognition of each other's efforts toward a common financial goal is a powerful reinforcement. Take time to praise your partner's smart shopping or the earning of an extra dollar.
- Make sure that you contribute enough income tax each month so that you are entitled to some sort of tax refund at the end of the year. Yes, this does mean you are lending money to the government interest free, but it builds in a nice reward for doing your taxes.
- Give yourself rewards for paying off debt or increasing your savings. This is easy-peasey if you have a definite budget, clear financial goals, and good record keeping.The numbers won't let you cheat, which is often the biggest problem with self-reward systems. Incorporate short-term as well as long-term payoffs, such as planning for dinners out and the eventual big vacation.

5. Let Your Passion Be Your Vocation

- Make your budget straightforward so you don't begrudge it. If you make your budget too restrictive, add too many layers to your financial plan, and insist on recording every pack of gum bought and consumed, you might have let the perfect get in the way of the good. The best financial plan is one you can stick with.
- Outsource part of your financial tasks, particularly taxes. Taxes are as certain as death and aren't something you can choose to ignore. If you get your taxes done on time, you avoid the instant late-filer penalty on the balance owed and the monthly late-filing payments that make matters worse. And most often you are

giving the taxman free money because you are actually owed a refund (i.e., typically $600 or more). Hiring a professional to complete your taxes for you will probably cost less than half of that. And if you are already a few years in arrears, there are tax specialists for that too. Let your problem be their passion.

IMPULSIVENESS AND DELAY

1. Precommitment and Self-Binding

- To prevent a wanton shopping spree, give yourself a reasonable allowance so that you can occasionally splurge. Entirely depriving yourself of a few treats ignores the principle of satiation. You want to satisfy underlying needs in a safe manner, or diminish them so they don't become uncontrollable.
- Put your money in a certificate of deposit or any other account that penalizes you for early withdrawal or keeps you from readily converting its contents to cash. It is the adult equivalent of a piggy bank—once we put the money in, it isn't so easy to get out.
- Make your credit card harder to access to give yourself a chance to have second thoughts. Double-sealing it in two Ziploc bags is helpful, especially if you have to pass through a copy of your last credit card statement that you keep in the first bag. A different approach is to leave it in your safe deposit box—or keep it on ice, literally. Freeze your credit card in a block of ice and store it in your refrigerator. Thaw it out when needed.
- Use preloaded credit cards like Green Dot. You load up the card with cash and can only spend what you put in. This also eliminates overdraft fees and penalty charges.
- In the extreme, cut up or otherwise destroy your credit cards. Do the same with all new credit card offerings *the moment* they arrive at your home.
- Go to your bank or your company's human resource department and commit to having them automatically increase your

tax payments, your deductions, or your savings *in the future*, say six months from now or next year. Sure, you will be able to undo these commitments, but that will require even more paperwork, which of course you will likely put off doing.

2. Making Paying Attention Pay

- Try paying with cash instead of credit cards, which tend to make the real cost of purchasing seem vague and surreal. There is nothing like laying down dollars to make you think twice.
- If you are using a credit card, at least take the time to review your monthly statement that shows you where the money goes. Try keeping it in your wallet so you can see it every time you go to make a purchase.
- If you really like a product, wait to buy it tomorrow or even next month—you'll most likely still like it then. Every business wants you to forgo comparison shopping and encourages impulsive purchases. But by waiting, not only do you avoid regretful purchases, you get the thrill of anticipation. Savoring actually increases the pleasure you receive from an item if you do ultimately purchase it. Wait for it.
- At tax time, find a cardboard box and print the word TAXES on it in bold letters. Leave it prominently displayed in a high-traffic area like the kitchen. Let the box cue you to fill it with all your tax receipts and forms.
- Have separate accounts for various purposes, such as savings, expenses, and splurges. This is called mental accounting, and works on the principle that we don't want to infect the serious money with the fun money by keeping both together. If you can separate them in your wallet, too, so much the better.
- Make your expenses tangible and as immediate as possible. For example, there is a wide selection of devices that move your electric meter indoors and put it in dollar figures (e.g., Control4, Kill

A Watt, Wattson Energy Meter, or the Electrisave). When you see your electricity usage shoot up like a taxi meter, you will find yourself naturally motivated to turn off a few more lights. Similarly, adding the ScanGauge to your car provides instantaneous feedback on cost per mile and cost per trip to help motivate you to drive more economically and curb your idling. In a pinch, you can tape your bills to prominent and relevant locations, such as the grocery bill to the fridge or the heating bill beside the thermostat.

3. Scoring Goals

- Make a series of short-term goals that are specific but challenging in size. Monthly or bi-weekly goals work quite well, especially if they are timed to your paychecks. In particular, tithing yourself is a popular technique that stretches back to Biblical times—take a tenth of your income and put it either toward debt or into savings.
- Chart a course for financial freedom. One of the most motivating ways of structuring debt is to service all your obligations but always focus on one source in particular. By setting up regular payments, you can schedule into your calendar when that targeted loan or purchase will be fully paid off. This gives you a nearby goal and, when it's reached, you can focus on your next source of debt. Even better, as you pay off one source of debt, the subsequent ones get paid down even faster. You'll like this; it creates an accelerated pace of repayment that we are psychologically hardwired to appreciate (i.e., a "debt snowball").
- Exploit the expertise of credit counselors—they are experts in creating a realistic but structured payment plan that will fit your budget.
- Set your own tax deadlines (before the official deadline, of course). We feel the most motivation as a deadline approaches. As always, be very specific. Try this one: "Immediately after

dinner tomorrow, I will sit in front of the computer and work on my taxes for one hour." Take a moment to visualize yourself doing this. Make that same goal for the following night and you will likely be done in a jiffy (remember, with tax software, most people are finished in under two hours).

- Plan your shopping trips ahead of time. By going in with a list, you avoid the impulse purchase. If something catches your eye, feel free to write down a few ideas for your next venture to the mall.
- Meet with your banker and set your bill payments and account deposits to occur automatically. And make use of programs like Quicken, which can load all your credit and debit card purchases at the click of a button. This makes budgeting that much easier. Normally, it takes weeks or months to develop a habit, to make actions happen without conscious effort. Fortunately, in finance we can cut down this timeline to a few minutes by automating a few things.
- Look at the fundraising "thermometers" that charities often use. They help make the abstract goal of raising cash a little more concrete, with definite milestones and progress reports. If you can create something similar, like a chart to financial freedom, and keep it prominently on your fridge, you will reap similar benefits regardless of whether you are paying down debt or increasing your savings.

LOOKING FORWARD

This is it. You are on your own now. This is as far as I can take you, at least in book form. I'm feeling a bit like I'm sending my kids off to their first day of school, but there will be less guilt if I know that I've left you prepared. I couldn't brownbag you a lunch, but I did give you the best of what I know, which should be more useful than a sandwich and a fruit cup. Do you understand why we put stuff off? Check. Do you know where the weak points are in your own motivational skills? Check. Do you have strategies for addressing these weak points? Check. And do you have some practical examples to help you implement them? Triple check. This should be enough to send you on your way. Just make sure you play nice and make friends.

Don't worry, though: you are going to do fine. Self-help books typically do help,[13] especially when the advice is scientifically backed. If you want a little more assistance, though, I won't take offense. Just as aspiring swimmers might want to employ a swim coach, hiring a life coach to walk you through a few of these techniques would be a wonderful complement. Another motivational book might give you additional insight by using another set of examples or by putting a greater emphasis on a particular subset of techniques. Do tell me if they bring up any new or useful principles, mostly because it would surprise me. To be effective, other experts will have to draw on the same sources I used. This isn't the guesswork of motivation—it is the science. This is what works.

ACKNOWLEDGMENTS

This book all started with a phone call from an immensely talented and likable literary agent, Sally Harding. After seeing my research covered in the press, she insisted long before anyone else that I was the person to write the book on procrastination. Who was I to argue? The Cooke Agency was wise to merge their agency with hers, and she equally so to form a partnership with them. With Dean Cooke, Suzanne Brandreth, and Mary Hu, they make a fine crew that can steer a book through any waters, foreign or domestic.

My thanks also go to Louise Dennys at the Knopf Random House Canada Group, who saw the potential of this book, and to the extraordinarily erudite Anne Collins, who wields a golden pen. She is an editor's editor, and becoming publisher at Knopf Random Canada was inevitable. Anne improved every page here. I am indebted to Nancy Miller, who championed the book early on and then to Jonathan Burnham at HarperCollins US, who ensured that it had a home. Also, I am grateful to my editor there, Sally Kim, for stubbornly insisting that what I thought was good enough should be better. Talented and thoughtful, she even gave me her own umbrella when I got caught in a New York rainstorm. Special thanks to Jane Isay for bringing the manuscript home by providing finishing editorial touches and making sure the narrative flowed. With her extensive experience and her familiarity with psychodynamics, psychology, and neurobiology, we made a good team. Lastly, the lovely Jane McWhinney gave the final polish, making sure each sentence gleamed. Like raising a child, writing a book takes a village, and I am thankful to have had so many gifted people in my corner.

Early in my academic career at the University of Minnesota, I was lucky to have Dr. Deniz Ones teach me meta-analysis and Dr. Thomas Brothen initiate my lifelong fascination with procrastination. At the University of Calgary, where I currently reside, much appreciation goes to my colleague and friend Dr. Daphne Taras, who fought to make sure I received my sabbatical to write this book and who provided, or credibly feigned, interest in the manuscript development. Though I wished the sabbatical had been longer, those uninterrupted months proved invaluable. I also appreciate the efforts of her son, Matthew Taras, for confirming historical facts. Further appreciation goes to my sisters, Anita and Marion, for reading earlier drafts and to my father-in-law, John Horne, a consulting economist, for his critical eye.

For everything else, and everything in general, I thank my wife, Julie. The conditions for writing this book, like so much of life, were not ideal and yet here it is. Teaching, researching, and running a department aren't easy for a parent of a toddler and a newborn. With both of our families in other cities, it seemed ridiculous to think I could also take on writing a book, but we did it anyway. My wife and I traded off sleeping on different nights, tag-teamed the children, and I absolutely relied on her support and faith. Though the motivational principles contained within this book proved invaluable, her reserves of strength are the platform on which this book was built. And, through all of it, I learned that she is a very gifted copy editor with a most discerning eye. The reader, as am I, should be very happy we are married.

NOTES

AUTHOR'S NOTE

1 Even the philosophers have been fascinated by procrastination, able to sit and watch it for hours:

Andreou, C. (2007). Understanding procrastination. *Journal for the Theory of Social Behavior, 37*(2), 183–193.

Gosling, J. (1990). *Weakness of the will.* New York: Routledge.

Silver, M. (1974). Procrastination. *Centerpoint, 1*(1), 49–54.

Sorensen, R. (2006). Originless sin: Rational dilemmas for satisficers. *The Philosophical Quarterly, 56*(223), 213–223.

2 Katz, I., de Deyn, P., Mintzer, J., Greenspan, A., Zhu, Y., & Brodaty, H. (2007). The efficacy and safety of risperidone in the treatment of psychosis of Alzheimer's disease and mixed dementia: A meta-analysis of 4 placebo-controlled clinical trials. *International Journal of Geriatric Psychiatry,* 22(5), 475–484.

Lee, J., Seto, D., & Bielory, L. (2008). Meta-analysis of clinical trials of probiotics for prevention and treatment of pediatric atopic dermatitis. *The Journal of Allergy and Clinical Immunology, 121*(1), 116–121.

3 Bowen, F., Rostami, M., & Steel, P. (2010). Meta-analysis of organizational innovation and performance. *Journal of Business Research.*

Caird, J., Willness, C. R., Steel, P., & Scialfa, C. (2008). A meta-analysis of the effects of cell phones on driver performance. *Accident Analysis & Prevention, 40*(4), 1282–1293.

Peloza, J., & Steel, P. (2005). The price elasticities of charitable contributions: A meta-analysis. *Journal of Public Policy & Marketing, 24*(2), 260–272.

Taras, V., Kirkman, B. L., & Steel, P. (2010). Examining the impact of Culture's Consequences: A three-decade, multi-level, meta-analytic review of Hofstede's cultural value dimensions. *Journal of Applied Psychology, 95* (3), 405–439.

Steel, P. & Kammeyer-Mueller, J. (2002). Comparing meta-analytic moderator search techniques under realistic conditions. *Journal of Applied Psychology, 87*(1), 96–111.

Steel, P., & Kammeyer-Mueller, J. (2008). Bayesian variance estimation for meta-analysis: Quantifying our uncertainty. *Organizational Research Methods, 11*(1), 54–78.

Steel, P., & Kammeyer-Mueller, J. (2009). Using a meta-analytic perspective to enhance Job Component Validation. *Personnel Psychology, 62*, 533–552.

Steel, P., & Ones, D. (2002). Personality and happiness: A national level of analysis. *Journal of Personality and Social Psychology, 83*(3), 767–781.

Steel, P., & Taras, V. (2010). Culture as a consequence: A multilevel multivariate meta-analysis of the effects of individual and country characteristics on work-related cultural values. *Journal of International Management.*

Steel, P., Schmidt, J., & Shultz, J. (2008). Refining the relationship between personality and subjective well-being. *Psychological Bulletin, 134*(1), 138–161.

CHAPTER ONE

1 Astrologers also cluster the twelve signs of the zodiac into quadruplicities, with the Gemini, Virgo, Sagittarius, and Pisces group

being particularly relevant. Quoting Bertrand Russell, this "quadruplicity drinks procrastination," as both Sagittarius and Pisces are inebriated with the trait. If you found that sentence awkward, I have a confession. In truth, Bertrand Russell actually intended these words to be an example of a sentence with correct grammar, but whose meaning is nonsensical. However, I was inspired to make sense out of nonsense by the Harvard linguist Yuen Ren Chao, who was Russell's interpreter when he went to China in the 1920s. He did the same for "Colorless green ideas sleep furiously." On the other hand, this type of wordplay doesn't increase your popularity.

2 Gendler, T. S. (2007). Self-deception as a pretense. *Philosophical Perspectives, 21*(1), 231–258.

Gosling, J. (1990). *Weakness of the will.* New York: Routledge.

Martin, M. (1986). *Self-deception and morality.* Lawrence, KS: University Press of Kansas.

3 Steel, P. (2007). The nature of procrastination: A meta-analytic and theoretical review of quintessential self-regulatory failure. *Psychological Bulletin, 133*(1), 65–94.

4 See www.43things.com, a website that has helped millions of people create their life lists.

5 Horn, S. (2001). *ConZentrate: Get focused and pay attention—when life is filled with pressures, distractions, and multiple priorities.* New York: Saint Martin's Press.

6 As part of my research, I documented the professions of 20,000 self-professed procrastinators. For example, it even afflicts pageant contestants; Sara Hoots, former Miss Hooters winner, revealed in her audition video: "My worst trait is procrastination." However, astronaut and zookeeper weren't among the ones I recorded. For those instances, we have confessions from *Slate*'s "Procrasti-Nation: Workers of the world, slack off!"

7 Gröpel, P., & Steel, P. (2008). A mega-trial investigation of goal-setting,

interest enhancement, and energy on procrastination. *Personality and Individual Differences, 45*, 406–411.

Silverman, I. (2003). Gender Differences in Delay of Gratification: A Meta-Analysis. *Sex Roles, 49*(9), 451–463.

8 Take your pick:

Burka, J. B., & Yuen, L. M. (1983). *Procrastination: Why you do it, what to do about it.* Reading, MA: Addison-Wesley.

Fiore, N. (1989). *The now habit: A strategic program for overcoming procrastination and enjoying guilt-free play.* New York: Penguin Putnam, Inc.

Knaus, W. (2002). *The procrastination workbook: Your personalized program for breaking free from the patterns that hold you back.* Oakland, CA: New Harbinger Publications, Inc.

Peterson, K. E. (1996). *The tomorrow trap: Unlocking the secrets of the procrastination-protection syndrome.* Deerfield Beach, FL: Health Communications, Inc.

9 McGarvey, J. (1996). The almost perfect definition. *Research/Penn State, 17*(3). Retrieved from http://www.rps.psu.edu/sep96/almost.html.

10 In addition to my "Nature of Procrastination" article, see also:

Canter, D. (2008). *Self-appraisals, perfectionism, and academics in college undergraduates.* Unpublished PhD, Virginia Commonwealth University, Richmond, VA.

Yao, M. (2009). *An exploration of multidimensional perfectionism, academic self-efficacy, procrastination frequency, and Asian American cultural values in Asian American university students.* Unpublished PhD, Ohio State University, Columbus, Ohio.

11 Pullen, F. J. (2003). *Perfectionism, procrastination, and other self-reported barriers to completing the doctoral dissertation.* Unpublished PhD, The University of Iowa, Iowa City, IA.

12 Schouwenburg, H. C. (2004). Academic procrastination: Theoretical notions, measurement, and research. In H. C. Schouwenburg,

C. H. Lay, T. A. Pychyl, & J. R. Ferrari (Eds.), *Counseling the procrastinator in academic settings* (pp. 3–17). Washington, DC: American Psychological Association.

13 Arce, E., & Santisteban, C. (2006). Impulsivity: A review. *Psicothema, 18*(2), 213–220.

Bembenutty, H., & Karabenick, S. A. (2004). Inherent association between academic delay of gratification, future time perspective, and self-regulated learning. *Educational Psychology Review, 16*(1), 35–57.

Enticott, P., & Ogloff, J. (2006). Elucidation of impulsivity. *Australian Psychologist, 41*(1), 3–14.

Whiteside, S., & Lynam, D. (2001). The Five Factor Model and impulsivity: Using a structural model of personality to understand impulsivity. *Personality and Individual Differences, 30*(4), 669–689.

14 Bui, N. H. (2007). Effect of evaluation threat on procrastination behavior. *Journal of Social Psychology, 147*(3), 197–209.

15 Schouwenburg, H. C. (2004). Academic procrastination: Theoretical notions, measurement, and research. In H. C. Schouwenburg, C. H. Lay, T. A. Pychyl, & J. R. Ferrari (Eds.), *Counseling the procrastinator in academic settings* (pp. 3–17). Washington, DC: American Psychological Association.

CHAPTER TWO

1 Overmier, J. B., & Seligman, M. E. P. (1967). Effects of inescapable shock upon subsequent escape and avoidance responding. *Journal of Comparative and Physiological Psychology, 63,* 28–33.

Seligman, M., & Csikszentmihalyi, M. (2000). Positive psychology: An introduction. *American Psychologist, 55,* 5–14.

Seligman, M. E. P., & Maier, S. F. (1967). Failure to escape traumatic shock. *Journal of Experimental Psychology, 74,* 1–9.

2 When I was taught about learned helplessness, my instructor related to me a story about a captured cricket. If you put a cricket in a jar,

give it some food and water, and punch in some small air holes for breathing, it will try to escape, launching itself up and hitting its head against the lid. Come back in a few days and take the lid off. The cricket will jump but just stop short of where the top used to be. It can escape at any time, but no longer engages in the behavior that will realize its freedom—the cage is now inside its head.

3 Beck, A. T., & Beck, R. W. (1972). Screening depressed patients in family practice: A rapid technique. *Postgraduate Medicine, 52*, 81–85.

4 Sadly, procrastination itself can even be a cause of the deviation-amplifying loop known as a depression spiral. That is, depression may lead to procrastination, which can cause guilt and self-loathing that deepens the depression, which completes the cycle by causing more procrastination. Such a hollowing out of existence can be further exacerbated if the activities one is putting off are of community and accomplishment, both of which help avoid depression from the start.

Thase, M. E. (1995). Cognitive behavior therapy. In I. D. Glick (Ed.), *Treating depression* (pp. 33–70). San Francisco: Jossey-Bass, Inc.

5 Lay, C. H. (1986). At last, my research article on procrastination. *Journal of Research in Personality, 20*(4), 474–495.

Lay, C. H. (1990). Working to schedule on personal projects: An assessment of person/project characteristics and trait procrastination. *Journal of Social Behavior & Personality, 5*(3), 9–103.

Milgram, N. (1988). Procrastination in daily living. *Psychological Reports, 63*(3), 752–754.

Milgram, N. A., Sroloff, B., & Rosenbaum, M. (1988). The procrastination of everyday life. *Journal of Research in Personality, 22*(2), 197–212.

Sirois, F. M. (2007). "I'll look after my health, later": A replication and extension of the procrastination-health model with community-dwelling adults. *Personality and Individual Differences, 43*(1), 15–26.

Sirois, F. M. (2007). Procrastination and motivations for household safety behaviors: An expectancy-value theory perspective. In L. V. Brown (Ed.), *Psychology of Motivation* (pp. 153–165): Nova Science Publishers.

6 Tullier, L. (2000). *The complete idiot's guide to overcoming procrastination.* Indianapolis, IN: Alpha Books.

7 Chainey, R. The death of the gym membership. Retrieved from http://style.uk.msn.com/getfit/sportandexercise/article.aspx?cp-documentid=9517875

8 Hershey, R. D. (November 28, 1999). Many shoppers won't do today what they can do on Dec. 24. *New York Times.*

9 Cosmides, L., & Tooby, J. (2000). Evolutionary psychology and the emotions. In M. Lewis & J. Haviland (Eds.), *Handbook of Emotions* (2 ed., pp. 91–115). New York: Guilford Press.

10 Whiteside, S., & Lynam, D. (2001). The Five Factor Model and impulsivity: Using a structural model of personality to understand impulsivity. *Personality and Individual Differences, 30*(4), 669–689.

11 McCrea, S., Liberman, N., Trope, Y., & Sherman, S. (2008). Construal level and procrastination. *Psychological Science, 19*(12), 1308–1314.

12 Here is Hume reflecting on how the nearby and concrete always seems to supersede the long-term and abstract: "In reflecting on any action which I am to perform a twelvemonth hence, I always resolve to prefer the greater good, whether at that time it will be more contiguous or remote; nor does any difference in that particular make a difference in my present intentions and resolutions. My distance from the final determination makes all those minute differences vanish, nor am I affected by anything but the general and more discernible qualities of good and evil. But on my nearer approach, those circumstances which I at first overlooked begin to appear, and have an influence on my conduct and affections. A new inclination to the present good springs up, and makes it difficult for me to adhere

inflexibly to my first purpose and resolution. This natural infirmity I may very much regret, and I may endeavour, by all possible means, to free myself from it."

13 Bagassi, M., & Macchi, L. (2007). The "vanishing" of the disjunction effect by sensible procrastination. *Mind & Society, 6*(1), 41–52.

14 Laven, A. V. (2007). *Freshmen college student mental health and their resource usage.* Unpublished EdD dissertation, University of California, Los Angeles, CA.

15 Cannings, R., Hawthorne, K., Hood, K., & Houston, H. (2005). Putting double marking to the test: A framework to assess if it is worth the trouble. *Medical Education, 39,* 299–308.

Newstead, S. (2002). Examining the examiners: Why are we so bad at assessing students? *Psychology Learning and Teaching, 2*(2), 70–75.

16 Caron, M. D., Whitbourne, S. K., & Halgan, R. P. (1992). Fraudulent excuse making among college students. *Teaching of Psychology, 19*(2), 90–93.

Lambert, E. G., Hogan, N. L., & Barton, S. M. (2003). Collegiate academic dishonesty revisited: What have they done, how often have they done it, who does it, and why did they do it? [Electronic Version]. *Electronic Journal of Sociology* 7. Retrieved July 11, 2008 from http://epe.lac-bac.gc.ca/100/201/300/ejofsociology/2004/v07n04/content/v017.4/lambert_etal.html.

Roig, M., & Caso, M. (2005). Lying and cheating: Fraudulent excuse making, cheating, and plagiarism. *The Journal of Psychology, 139*(6), 485–494.

Roig, M., & DeTommaso, L. (1995). Are college cheating and plagiarism related to academic procrastination? *Psychological Reports,* 77(2), 691–698.

17 The graph includes two-thirds of the courses students, excluding those who dropped the course or had finished the entire course work more than four days before the deadline and couldn't be potentially

procrastinating in this part of their lives. See also, the following articles that find procrastination almost perfectly fits a hyperbolic curve.

Green, L., & Myerson, J. (2004). A discounting framework for choice with delayed and probabilistic rewards. *Psychological Bulletin, 130*(5), 769–792.

Howell, A. J., Watson, D. C., Powell, R. A., & Buro, K. (2006). Academic procrastination: The pattern and correlates of behavioral postponement. *Personality and Individual Differences, 40*(8), 1519–1530.

Schouwenburg, H. C., & Groenewoud, J. T. (2001). Study motivation under social temptation: Effects of trait procrastination. *Personality & Individual Differences, 30*(2), 229–240.

CHAPTER THREE

1 Schelling, T. C. (1984). *Choice and consequence. Perspectives of an errant economist.* Cambridge: Harvard University Press.

2 Baumeister, R. (2005). *The cultural animal.* New York: Oxford University Press.

Bazerman, M. H., Tenbrunsel, A. E., & Wade-Benzoni, K. (1998). Negotiating with yourself and losing: Making decisions with competing internal preferences. *The Academy of Management Review,* 23(2), 225–241.

Bechara, A. (2005). Decision making, impulse control and loss of willpower to resist drugs: A neurocognitive perspective. *Nature Neuroscience, 8,* 1458–1463.

Bernheim, D., & Rangel, A. (2002). *Addiction, cognition, and the visceral brain.* Mimeo: Stanford University.

Chaiken, S., & Trope, Y. (1999). *Dual-process theories in social psychology.* New York: Guilford Press.

Loewenstein, G., & O'Donoghue, T. E. D. (2005). *Animal spirits: Affective and deliberative processes in economic behavior.* Carnegie Mellon University.

Metcalfe, J., & Mischel, W. (1999). A hot/cool-system analysis of delay of gratification: Dynamics of willpower. *Psychological Review, 106*(1), 3–19.

Redish, A., Jensen, S., & Johnson, A. (2008). A unified framework for addiction: Vulnerabilities in the decision process. *Behavioral and Brain Sciences, 31*(4), 415–437.

Sanfey, A. G., Loewenstein, G., McClure, S. M., & Cohen, J. D. (2006). Neuroeconomics: Cross-currents in research on decision-making. *TRENDS in Cognitive Sciences, 10*(3), 108–116.

3 As William James, the godfather of psychology, puts it when discussing the economic theory of behavior: "Not one man in a billion, when taking his dinner, ever thinks of utility. He eats because the food tastes good and makes him want more."

4 Hariri, A. R., Brown, S. M., Williamson, D. E., Flory, J. D., Wit, H. D., & Manuck, S. B. (2006). Preference for immediate over delayed rewards is associated with magnitude of ventral striatal activity. *The Journal of Neuroscience, 26*(51), 13213–13217.

McClure, S. M., Ericson, K. M., Laibson, D. I., Loewenstein, G., & Cohen, J. D. (2007). Time discounting for primary rewards. *Journal of Neuroscience, 27*(21), 5796–5804.

McClure, S. M., Laibson, D. I., Loewenstein, G., & Cohen, J. D. (2004). Separate neural systems value immediate and delayed monetary rewards. *Science, 306*(5695), 503–507.

5 Ainslie, G., & Monterosso, J. (2004). A marketplace in the brain? *Science, 306,* 421–423.

Banich, M. T. (2009). Executive function: The search for an integrated account. *Current Directions in Psychological Science, 18*(2), 89–94.

Bechara, A. (2005). Decision making, impulse control and loss of willpower to resist drugs: A neurocognitive perspective. *Nature Neuroscience, 8,* 1458–1463.

Rudebeck, P. H., Walton, M. E., Smyth, A. N., Bannerman, D. M., &

Rushworth, M. F. S. (2006). Separate neural pathways process different decision costs. *Nature Neuroscience, 9*(9), 1161–1168.

Spinella, M., Yang, B., & Lester, D. (2004). Prefrontal system dysfunction and credit card debt. *International Journal of Neuroscience, 114,* 1323–1332.

Walton, M. E., Rudebeck, P. H., Bannerman, D. M., & Rushworth, M. F. S. (2007). Calculating the cost of acting in frontal cortex. *Annals of the New York Academy of Sciences, 1104,* 340–356.

Wood, J. N., & Grafman, J. (2003). Human prefrontal cortex: Processing and representational perspectives. *Nature Reviews, 4,* 139–147.

6 Carver, C., Johnson, S., & Joormann, J. (2008). Serotonergic function, two-mode models of self-regulation, and vulnerability to depression: What depression has in common with impulsive aggression. *Psychological Bulletin, 134*(6), 912–943.

Fudenberg, D., & Levine, D. (2006). A dual-self model of impulse control. *American Economic Review, 96*(5), 1449–1476.

Inbinder, F. C. (2006). Psychodynamics and executive dysfunction: A neurobiological perspective. *Clinical Social Work Journal, 34*(4), 515–529.

Marcus, G. (2008). *Kluge: The haphazard construction of the human mind.* New York: Houghton Mifflin Company.

7 As Adam Gifford puts it:

> Evolution cannot discard existing designs and start over from scratch, it can only build the new on top of the old—the old higher biology-based time preference mechanisms are still built into the human brain. These mechanisms must be overridden in decision making by the inhibition process, which is significantly enhanced in humans by language. It is this divergence between the cultural and biological rates of time preference that creates a potential internal nature versus nurture conflict leading to self-control problems [like procrastination]. The higher level

> prefrontal working memory system allows the agent to consider possible events in the extended future and to discount those events at a rate appropriate to the individual's current environment. The lower level [limbic system] does not have access to events not yet experienced, and as a result, ignores these purely abstract events; it also incorporates the high level discount rate similar to that used by non-human primates and some other mammals that is a product of natural selection.

Gifford, A. (2002). Emotion and self-control. *Journal of Economic Behavior & Organization, 49,* 113–130.

8 Damasio, A. R. (1994). *Descartes' error: Emotion, reason, and the human brain.* New York: G.P. Putnam.

Gifford, A. (2002). Emotion and self-control. *Journal of Economic Behavior & Organization, 49,* 113–130.

McCrea, S. M., Liberman, N., Trope, Y., & Sherman, S. J. (2008). Construal level and procrastination. *Psychological Science 19*(12), 1308–1314.

Trope, Y., & Liberman, N. (2003). Temporal construal. *Psychological Review, 110*(3), 403–421.

Wood, J. N., & Grafman, J. (2003). Human prefrontal cortex: Processing and representational perspectives. *Nature Reviews, 4,* 139–147.

9 Berns, G. S., Laibson, D., & Loewenstein, G. (2007). Intertemporal choice—toward an integrative framework. *TRENDS in Cognitive Sciences, 11*(11), 482–488.

10 Brown, T. E. (2000). Emerging understandings of attention-deficit disorders and comorbidities. In T. E. Brown (Ed.), *Attention-deficit disorders and comorbidities in children, adolescents, and adults* (pp. 3–55). Washington, DC: American Psychiatric.

Reyna, V. F., & Farley, F. (2006). Risk and rationality in adolescent decision making: Implications for theory, practice, and public policy. *Psychological Science in the Public Interest* 7(1), 1–44.

Rosati, A. G., Stevens, J. R., Hare, B., & Hauser, M. D. (2007). The evolutionary origins of human patience: temporal preferences in chimpanzees, bonobos, and human adults. *Current Biology, 17*(19), 1663–1668.

Rosso, I. M., Young, A. D., Femia, L. A. & Yurgelun-Todd, D. A. (2004). Cognitive and emotional components of frontal lobe functioning in childhood and adolescence. *Annals of the New York Academy of Sciences, 1021,* 355–362.

Rubia, K., Overmeyer, S., Taylor, E., Brammer, M., Williams, S. C. R., Simmons, A., et al. (1999). Hypofrontality in Attention Deficit Hyperactivity Disorder during higher-order motor control: A study with functional MRI. *American Journal of Psychiatry, 156*(6), 891–896.

Stevens, J. R., Hallinan, E. V., & Hauser, M. D. (2005). The ecology and evolution of patience in two New World primates. *Biology Letters, 1,* 223–226.

Wood, J. N., & Grafman, J. (2003). Human prefrontal cortex: Processing and representational perspectives. *Nature Reviews, 4,* 139–147.

Yurgelun-Todd, D. A. & Killgore, W. D. S. (2006) Fear-related activity in the prefrontal cortex increases with age during adolescence: A preliminary fMRI study. *Neuroscience Letters, 406,* 194–199.

11 Miller B. L., Seeley, W. W., Mychack, P., Rosen, H. J., Mena, I., & Boone, K. (2001). Neuroanatomy of the self: Evidence from patients with frontotemporal dementia. *Neurology, 57,* 817–821.

12 Heilman, K. (2002). *Matter of mind: A neurologist's view of the brain-behavior relationships*. Oxford: Oxford University Press.

13 Knoch, D., & Fehr, E. (2007). Resisting the power of temptations: The right prefrontal cortex and self-control. *Annals of the New York Academy of Sciences, 1104,* 123–134.

14 Bechara, A. (2005). Decision making, impulse control and loss of willpower to resist drugs: A neurocognitive perspective. *Nature Neuroscience, 8,* 1458–1463.

Bickel, W. K., Miller, M. L., Yi, R., Kowal, B. P., Lindquist, D. M., & Pitcock, J. A. (2007). Behavioral and neuroeconomics of drug addiction: Competing neural systems and temporal discounting processes. *Drug and Alcohol Dependence, 90,* 85–91.

Gifford, A. (2002). Emotion and self-control. *Journal of Economic Behavior & Organization, 49,* 113–130.

15 Camerer, C., Loewenstein, G., & Prelec, D. (2005). Neuroeconomics: How neuroscience can inform economics. *Journal of Economic Literature, 43*(1), 9–64.

Joireman, J., Balliet, D., Sprott, D., Spangenberg, E., & Schultz, J. (2008). Consideration of future consequences, ego-depletion, and self-control: Support for distinguishing between CFC-Immediate and CFC-Future sub-scales. *Personality and Individual Differences, 45*(1), 15–21.

16 Reyna, V. F., & Farley, F. (2006). Risk and rationality in adolescent decision making: Implications for theory, practice, and public policy. *Psychological Science in the Public Interest* 7(1), 1–44.

Rosso, I. M., Young, A. D., Femia, L. A. & Yurgelun-Todd, D. A. (2004). Cognitive and emotional components of frontal lobe functioning in childhood and adolescence. *Annals of the New York Academy of Sciences, 1021,* 355–362.

Wood, J. N., & Grafman, J. (2003). Human prefrontal cortex: Processing and representational perspectives. *Nature Reviews, 4,* 139–147.

Yurgelun-Todd, D. A. & Killgore, W. D. S. (2006) Fear-related activity in the prefrontal cortex increases with age during adolescence: A preliminary fMRI study. *Neuroscience Letters, 406,* 194–199.

17 Thompson-Schill, S. L., Ramscar, M., & Chrysikou, E. G. (2009). Cognition without control: When a little frontal lobe goes a long way. *Current Directions in Psychological Science, 18*(5), 259–263.

18 Garon, N., Bryson, S., & Smith, I. (2008). Executive function in preschoolers: A review using an integrative framework. *Psychological Bulletin, 134*(1), 31–60.

Jurado, M., & Rosselli, M. (2007). The elusive nature of executive functions: A review of our current understanding. *Neuropsychology Review, 17*(3), 213–233.

19 Reyna, V. F., & Farley, F. (2006). Risk and rationality in adolescent decision making: Implications for theory, practice, and public policy. *Psychological Science in the Public Interest 7*(1), 1–44.

20 Jurado, M., & Rosselli, M. (2007). The elusive nature of executive functions: A review of our current understanding. *Neuropsychology Review, 17*(3), 213–233.

21 Miller B. L., Seeley, W. W., Mychack, P., Rosen, H. J., Mena, I., & Boone, K. (2001). Neuroanatomy of the self: Evidence from patients with frontotemporal dementia. *Neurology, 57,* 817–821.

22 Dingemanse, N., & Réale, D. (2005). Natural selection and animal personality. *Behaviour, 142*(9), 1159–1184.

23 Gosling, S., Kwan, V., & John, O. (2003). A dog's got personality: A cross-species comparative approach to personality judgments in dogs and humans. *Journal of Personality and Social Psychology, 85*(6), 1161–1169.

24 Mazur, J. (2001). Hyperbolic value addition and general models of animal choice. *Psychological Review, 108*(1), 96–112.

Stephens, D. W., Kerr, B., & Fernandez-Juricic, E. (2004). Impulsiveness without discounting: The ecological rationality hypothesis. *Proceedings—Royal Society of London: Biological sciences 271,* 2459–2465.

Stuphorn, V. (2005). Neuroeconomics: The shadow of the future. *Current Biology, 15*(7), 247–249.

25 Suddendorf, T., & Corballis, M. C. (2007). The evolution of foresight: What is mental time travel and is it unique to humans? *Behavioral and Brain Sciences, 30*(3), 299–351.

Roberts, W. A. (2007). Mental time travel: Animals anticipate the future. *Current Biology, 17*(11), R418–R420.

26 Roberts, W. A., Feeney, M. C., MacPherson, K., Petter, M., McMillan, N., & Musolino, E. (2008). Episodic-like memory in rats: Is it based on when or how long ago? *Science, 320*(5872), 113–115.

27 Mischel, W., & Ayduk, O. (2004). Willpower in a cognitive-affective processing system. In I. Baumeister & K. Vohs (Eds.), *Handbook of self-regulation: Research, theory, and applications* (pp. 99–129). New York: Guilford Press.

Rosati, A. G., Stevens, J. R., Hare, B., & Hauser, M. D. (2007). The evolutionary origins of human patience: Temporal preferences in chimpanzees, bonobos, and human adults. *Current Biology, 17*(19), 1663–1668.

Stevens, J. R., Hallinan, E. V., & Hauser, M. D. (2005). The ecology and evolution of patience in two New World primates. *Biology Letters, 1,* 223–226.

28 Gomes, C. M., & Boesch, C. (2009). Wild chimpanzees exchange meat for sex on a long-term basis. *PLoS ONE, 4*(4), 5116.

29 Osvath, M. (2009). Spontaneous planning for future stone throwing by a male chimpanzee. *Current Biology, 19*(5), R190–R191.

30 Ainslie, G. (1974). Impulse control in pigeons. *Journal of the Experimental Analysis of Behavior, 21*(3), 485-489.

Biondi, D. R. (2007). *Procrastination in rats: The effect of delay on response requirements in an adjusting ratio procedure.* Unpublished M.A. dissertation, Southern Connecticut State University, New Haven, CT.

Mazur, J. E. (1996). Procrastination by pigeons: Preferences for larger, more delayed work requirements. *Journal of the Experimental Analysis of Behavior, 65*(1), 159–171.

Mazur, J. E. (1998). Procrastination by pigeons with fixed-interval response requirements. *Journal of the Experimental Analysis of Behavior, 69*(2), 185–197.

Rachlin, H., & Green, L. (1972). Commitment, choice and self-control. *Journal of the Experimental Analysis of Behavior, 17*(1), 15-22.

31 Indeed, the reason why pigeons can procrastinate is that they do have a prefrontal cortex counterpart, the *nidopallium caudolaterale* .

Güntürkün, O. (2005). The avian 'prefrontal cortex' and cognition. *Current Opinion in Neurobiology, 15*(6), 686–693.

32 As Cesar Millan stressed, to instill discipline in your pet, you need to have discipline in yourself. "Exercise, discipline, and affection." Too often the middle ingredient is left out.

Arden, A., & Dockray, T. (2007). *Dog-friendly dog training* (2nd ed.). New York: John Wiley and Sons.

33 Jang, K. L., McCrae, R. R., Angleitner, A., Riemann, R., & Livesley, W. J. (1998). Heritability of facet-level traits in a cross-cultural twin sample: Support for a hierarchical model of personality. *Journal of Personality and Social Psychology, 74*(6), 1556–1565.

Luciano, M., Wainwright, M. A., Wright, M. J., & Martin, N. G. (2006). The heritability of conscientiousness facets and their relationship to IQ and academic achievement. *Personality and Individual Differences, 40,* 1189–1199.

Notably, this finding is consistent with other personality studies, which typically estimate that between 40 and 60 percent of any personality trait is genetic in origin. Bouchard, T., & Loehlin, J. (2001). Genes, evolution, and personality. *Behavior Genetics, 31*(3), 243–273.

34 Dingemanse, N., & Réale, D. (2005). Natural selection and animal personality. *Behaviour, 142*(9), 1159–1184.

Sih, A., Bell, A., & Johnson, J. (2004). Behavioral syndromes: An ecological and evolutionary overview. *Trends in Ecology & Evolution, 19*(7), 372–378.

35 Whit, W. (1995). *Food and society: A sociological approach.* Dix Hills, NY: General Hall.

36 Stevens, J. R., Hallinan, E. V., & Hauser, M. D. (2005). The ecology and evolution of patience in two New World primates. *Biology Letters, 1,* 223–226.

37 Houston, A. I., McNamara, J. M., & Steer, M. D. (2007). Do we expect natural selection to produce rational behaviour? *Philosophical Transactions of the Royal Society B: Biological Sciences, 362,* 1531–1543.

38 Kalenscher, T., & Pennartz, C. M. A. (2008). Is a bird in the hand worth two in the future? The neuroeconomics of intertemporal decision-making. *Progress in Neurobiology, 84*(3), 284–315.

39 Davies, D. W. (1983). *Owen Owen: Victorian draper.* Gwasg Cambria: Aberystwyth.

There is also a Wikipedia page: http://en.wikipedia.org/wiki/Owen_Owen

40 Schmitt, D. (2004). The Big Five related to risky sexual behaviour across 10 world regions: Differential personality associations of sexual promiscuity and relationship infidelity. *European Journal of Personality, 18*(4), 301–319.

Raffaelli, M., & Crockett, L. (2003). Sexual risk taking in adolescence: The role of self-regulation and attraction to risk. *Developmental Psychology, 39*(6), 1036–1046.

Reyna, V. F., & Farley, F. (2006). Risk and rationality in adolescent decision making: Implications for theory, practice, and public policy. *Psychological Science in the Public Interest* 7(1), 1–44.

41 Silverman, I. (2003). Gender differences in delay of gratification: A meta-analysis. *Sex Roles, 49*(9), 451–463.

42 Nettle, D. (2006). The evolution of personality variation in humans and other animals. *American Psychologist, 61*(6), 622–631.

Muller, H., & Chittka, L. (2008). Animal personalities: The advantage of diversity. *Current Biology, 18*(20), R961–R963.

Nichols, C. P., Sheldon, K. M., & Sheldon, M. S. (2008). Evolution and personality: What should a comprehensive theory address and how? *Social and Personality Psychology Compass,* 2(2), 968–984.

Planque, R., Dornhaus, A., Franks, N. R., Kovacs, T., & Marshall, J. A. R. (2007). Weighting waiting in collective decision-making. *Behavioral Ecology and Sociobiology, 61*(3), 347–356.

43 Smith, E., Mulder, M., & Hill, K. (2001). Controversies in the evolutionary social sciences: A guide for the perplexed. *Trends in Ecology & Evolution, 16*(3), 128–135.

44 To investigate this topic, I tried to locate a 1971 book by Paul T. Ringenbach, *Procrastination through the Ages: A Definitive History.* Ringenbach is a U.S. Air Force Officer with a PhD from the University of Connecticut. His work was described as "an interesting survey" by the late Albert Ellis on the very first page of his book *Overcoming Procrastination,* making it a must-have for anyone interested in the topic. After spending weeks hunting with a pack of librarians, I finally found some correspondence with Gil Campbell from *Filter Press,* the book's publisher, buried in the appendix of an old 1982 doctoral thesis by Margaret Aitken. The letter indicated that *Procrastination through the Ages* was never actually written. Colonel Ringenbach was asked to write it, but kept putting it off for so long that it metamorphosized into an elaborate prank, with Campbell telling everyone for fifteen years that it was coming out imminently. I tracked Colonel Ringenbach to his Texan address, where after a series of e-mails and phone calls, I extracted a full confession.

> By the speed of my response, I guess you realize that procrastination is alive and well. *Procrastination through the Ages: A Definitive History* first appeared in Books in Print in the 1971–1972 edition. How it came about was that Gil Campbell of the Filter Press was also the Acquisitions chief at the US Air Force Academy when I first met him . . . He asked me to do a short piece for him on "Black Cowboys" that he could publish. After a time with no progress, he suggested I write a book on procrastination because I was so good about it. Months rolled on with no progress, so finally he said give me a title; I want to include it in my next catalogue. I gave him the title and he did not print it in the catalogue on purpose but included it as a loose insert on colored paper with the excuse to the readers that he hadn't got around to including it in the actual text, but here it is anyway. At this

point he put it in Books in Print with a date not set, price not set notation. After all we surmised, how could one ever complete a book on procrastination? It continued in Books in Print about 15 years until Gil took it out because he was tiring of the continuing inquiries that he always sent along dutifully to me to reply.

45 DeSimone, P. (1993). Linguistic assumptions in scientific language. *Contemporary Psychodynamics: Theory, Research & Application, 1*, 8–17. Of note, a firsthand copy of DeSimone's work is no longer in existence. The journal in which it appeared, *Contemporary Psychodynamics,* had a run of but a single issue and no copies have so far been obtainable. DeSimone's work is reviewed in the book *Procrastination and Task Avoidance: Theory, Research, and Treatment,* which I am using as a proxy.

46 Handily available in book form as well, called the *Phillipics*.

47 Olcott, H. S. (1887). *Golden rules of Buddhism*. London: Theosophical Publishing House.

48 Ziolkowski, T. (2000). *The sin of knowledge: Ancient themes and modern variations*. Princeton: Princeton University Press.

49 Diamond, J. (1987, May). The worst mistake in the history of the human race. *Discover,* 64–66.

CHAPTER FOUR

1 Though officially not being "associated with Risk or Hasbro in any way."

2 Steel, P. (2002). *The measurement and nature of procrastination*. Unpublished PhD dissertation, University of Minnesota, Minnesota, MN.

3 Schlinger, H. D., Derenne, A., & Baron, A. (2008). What 50 years of research tell us about pausing under ratio schedules of reinforcement. *The Behavior Analyst, 31*, 39–40.

4 Czerny, E., Koenig, S., & Turner, N. E. (2008). Exploring the mind of the gambler: Psychological aspects of gambling and problem

gambling In M. Zangeneh, A. Blaszczynski & N. Turner (Eds.), *In the pursuit of winning* (pp. 65–82). New York: Springer.

5 Latham, G., & Huber, V. (1992). Schedules of reinforcement: Lessons from the past and issues for the future. *Journal of Organizational Behavior Management, 12*(1), 125–149.

6 Taras, V., & Steel, P. (2006). *Improving cultural indices and rankings based on a meta-analysis of Hofstede's taxonomy.* Paper presented at the Academy of International Business Annual Meeting, Beijing, China. Best paper in Managing People across Border track and nominated for AIB Best Paper/Temple award (overall conference best).

Steel, P. (2007). The nature of procrastination. *Psychological Bulletin, 133*(1), 65–94.

7 Pelman Institute of America (March, 1930). The man with the grasshopper mind. *Popular Mechanics, 53*(3), 336.

8 Josephs, R. (January, 1962) How to gain an extra hour every day. *Popular Science, 180*(1), 117–130.

9 Myers, D. G. (1983) *Social psychology.* New York: McGraw-Hill.

10 Glater, J. D. (2008). Welcome, freshmen. Have an iPod. *New York Times*. Retrieved from: http://www.nytimes.com/2008/08/21/technology/21iphone.html?th&emc=th

11 Pychyl, T. A., Lee, J. M., Thibodeau, R., & Blunt, A. (2000). Five days of emotion: An experience sampling study of undergraduate student procrastination. *Journal of Social Behavior & Personality, 15*(5), 239–254.

12 Frey, B. S., Benesch, C., & Stutzer, A. (2007). Does watching TV make us happy? *Journal of Economic Psychology, 28*(3), 283–313.

13 Kubey, R., & Csikszentmihalyi, M. (2002). Television addiction is no mere metaphor. *Scientific American, 286*(2), 62–68.

Vandewater, E., Bickham, D., & Lee, J. (2006). Time well spent? Relating television use to children's free-time activities. *Pediatrics, 117*(2), 181–191.

14 Harchandrai, P., & Whitney, J. (2006). *Video games are cooler than homework: The role of video games in procrastination.* Paper presented at the Conference for Undergraduate Research in Communication, Rochester Institute of Technology.

15 Applebome, P. (2004, December 1, 2004). On campus, hanging out by logging on. *New York Times.*

16 Aspan, M. (February 13, 2008). Quitting Facebook gets easier. *New York Times.*

17 Kessler, D. A. (2009). *The end of overeating: Taking control of the insatiable American appetite.* New York: Rodale.

18 Offer, A. (2006). *The challenge of affluence: Self-control and well-being in the United States and Britain since 1950.* New York: Oxford University Press.

19 Dittmar, H. (2005). Compulsive buying—a growing concern? An examination of gender, age, and endorsement of materialistic values as predictors. *British Journal of Psychology, 96,* 467–491.

LaRose, R., & Eastin, M. S. (2002). Is online buying out of control? Electronic commerce and consumer self-regulation. *Journal of Broadcasting and Electronic Media, 46*(4), 549–564.

Percoco, M. (2009). Estimating individual rates of discount: A meta-analysis. *Applied Economics Letters, 6*(12), 1235–1239.

Verplanken, B., & Herabadi, A. (2001). Individual differences in impulse buying tendency: Feeling and no thinking. *European Journal of Personality, 15,* 71–83.

Youn, S., & Faber, R. J. (2000). Impulse buying: Its relation to personality traits and cues. *Advances in Consumer Research, 27,* 179–185.

20 Baumeister, R. F. (2002). Yielding to temptation: Self-control failure, impulsive purchasing, and consumer behavior. *Journal of Consumer Research, 28,* 670–676.

Baumeister, R., Sparks, E., Stillman, T., & Vohs, K. (2008). Free will in consumer behavior: Rational choice and self-control. *Journal of Consumer Psychology, 18,* 4–13.

LaRose, R., & Eastin, M. S. (2002). Is online buying out of control? Electronic commerce and consumer self-regulation. *Journal of Broadcasting and Electronic Media, 46*(4), 549–564.

Lynch, J. G., & Zauberman, G. (2006). When do you want it? Time, decisions, and public policy. *Journal of Public Policy & Marketing, 25*(1), 67–78.

Ziglar, Z. (1991). *Ziglar on selling.* New York: Thomas Nelson.

21 Kessler, D. A. (2009). *The end of overeating: Taking control of the insatiable American appetite.* New York: Rodale.

22 Duhigg, C. (July 13, 2008). Warning: Habits may be good for you. *New York Times.*

23 Ji, M., & Wood, W. (2007). Purchase and consumption habits: Not necessarily what you intend. *Journal of Consumer Psychology, 17*(4), 261–276.

24 Wood, W., & Neal, D. T. (2007). A new look at habits and the habit-goal interface. *Psychological Review, 114*(4), 843–863.

25 Wansink, B. (2006). *Mindless eating: Why we eat more than we think.* New York: Bantam-Dell.

26 Ariely, D., Loewenstein, G., & Prelec, D. (2006). Tom Sawyer and the construction of value. *Journal of Economic Behavior & Organization, 60*(1), 1–10.

Lindstrom, M. (2005). *BRAND sense: Build powerful brands through touch, taste, smell, sight, and sound.* New York: Free Press.

Ramanathan, S., & Menon, G. (2006). Time-varying effects of chronic hedonic goals on impulsive behavior. *Journal of Marketing Research, 43*(4), 628–641.

Wood, W., & Neal, D. T. (2007). A new look at habits and the habit-goal interface. *Psychological Review, 114*(4), 843–863.

27 Caird, J., Willness, C. R., Steel, P., & Scialfa, C. (2008). A meta-analysis of the effects of cell phones on driver performance. *Accident Analysis & Prevention, 40*(4), 1282–1293.

28 Strictly speaking, there were a few other categories but all in the

same genre, such as *Shazam* for "Music" or the *Virtual Zippo Lighter* for "Lifestyle."

29 Huxley, A. (2004). *Brave New World and Brave New World Revisited.* New York: HarperCollins.

30 Postman, N. (1985). *Amusing ourselves to death: Public discourse in the age of show business.* New York: Penguin Group.

31 Offer, A. (2006). *The challenge of affluence: Self-control and well-being in the United States and Britain since 1950.* New York: Oxford University Press.

Novotney, A. (July/August, 2008). What'$ behind American con$umeri$m? *Monitor on Psychology, 39*(7), 40–42.

Vyse, S. (2008). *Going broke: Why Americans can't hold on to their money.* New York: Oxford University Press.

32 Davenport, T., & Beck, J. (2001). *The Attention Economy: Understanding the new currency of business.* Harvard Business School Press.

Shenk, D. (1997). *Data smog: Surviving the information glut.* New York: HarperCollins.

CHAPTER FIVE

1 Ferrari, J. R., Barnes, K. L., & Steel, P. (2009). Life regrets by avoidant and arousal procrastinators: Why put off today what you will regret tomorrow? *Journal of Individual Differences, 30*(3), 163–168.

Roese, N. J., & Summerville, A. (2005). What we regret most . . . and why. *Personality and Social Psychology Bulletin, 31*(9), 1273–1285.

2 Steel, P., Schmidt, J., & Shultz, J. (2008). Refining the relationship between personality and subjective well-being. *Psychological Bulletin, 134*(1), 138–161.

3 Baer, M., & Oldham, G. R. (2006). The curvilinear relation between experienced creative time pressure and creativity: Moderating effects

of openness to experience and support for creativity. *Journal of Applied Psychology, 91,* 963–970.

Amabile, T. M., Hadley, C. N., & Kramer, S. J. (2002). Creativity under the gun. *Harvard Business Review, 80*(8), 52–61.

4 Steel, P. (2007). The nature of procrastination: A meta-analytic and theoretical review of quintessential self-regulatory failure. *Psychological Bulletin, 133*(1), 65–94.

5 Pychyl, T. A., Lee, J. M., Thibodeau, R., & Blunt, A. (2000). Five days of emotion: An experience-sampling study of undergraduate student procrastination. *Journal of Social Behavior & Personality, 15*(5), 239–254.

6 Patry, D. A., Blanchard, C. M., & Mask, L. (2007). Measuring university students' regulatory leisure coping styles: planned breathers or avoidance? *Leisure Sciences, 29*(3), 247–265.

7 Bernold, L. E. (2007). Preparedness of engineering freshman to inquiry-based learning. *Journal of Professional Issues in Engineering Education and Practice, 133,* 99–106.

Doherty, W. (2006). An analysis of multiple factors affecting retention in Web-based community college courses. *The Internet and Higher Education, 9*(4), 245–255.

Finck, J., & DeLine, A. (2008). Do students listen to advice from their experienced peers? *College Teaching Methods & Styles Journal, 4*(9), 19–26.

Laven, A. V. (2007). *Freshmen college student mental health and their resource usage.* Unpublished EdD dissertation, University of California, Los Angeles, CA.

Moore, B. (2006). *Goal conflicts, self-regulation, and course completion: A comparison of Web-based learners to traditional classroom learners.* Unpublished PhD dissertation, University of South Florida, Tampa, FL.

8 Bair, C. R., & Haworth, J. G. (2004). Doctoral student attrition and persistence: A meta-synthesis of research. *Higher education: Handbook of theory and research, 19,* 481–534.

Green, G. D. (1981). *Dissertation procrastination.* Unpublished PhD dissertation, University of Washington, Seattle, WA.

Muszynski, S. Y., & Akamatsu, T. J. (1991). Delay in completion of doctoral dissertations in clinical psychology. *Professional Psychology—Research & Practice,* 22(2), 119–123.

Mariano, C. M. (1993). *A study of Ed.D.s, Ph.D.s and ABDs in educational administration (dissertation completion, Ed.D. candidates, Ph.D. candidates).* Unpublished EdD dissertation, Boston College, Boston, MA.

Pullen, F. J. (2003). *Perfectionism, procrastination, and other self-reported barriers to completing the doctoral dissertation.* Unpublished PhD dissertation, University of Iowa, New Haven, IA.

9 Based on the average salary difference between those with a Master's and those with a Doctorate education.

Lacey, J. & Crosby, O. (2005). Job outlook for college graduates. *Occupational Outlook Quarterly, 48*(4), 15–27.

10 Lay, C. H., & Brokenshire, R. (1997). Conscientiousness, procrastination, and person-task characteristics in job searching by unemployed adults. *Current Psychology: Developmental, Learning, Personality, Social, 16*(1), 83–96.

Senecal, C., & Guay, F. (2000). Procrastination in job-seeking: An analysis of motivational processes and feelings of hopelessness. *Journal of Social Behavior & Personality, 15*(5), 267–282.

11 Nawrocki, J. (2006, June 15, 2006). When you're a GC, procrastination doesn't work. *Corporate Counsel,* Retrieved from http://www.law.com/jsp/ihc/PubArticleIHC.jsp?id=1150275918375

12 Angeletos, G.-M., Laibson, D., Repetto, A., Tobacman, J., & Weinberg, S. (2001). The hyperbolic consumption model: Calibration, simulation, and empirical evaluation. *Journal of Economic Perspectives, 15*(3), 47–68.

13 Bankston, J. (2001). IRS experts blame procrastination for simple oversights on tax returns. *The Augusta Chronicle, GA. Knight Ridder/Tribune Business News.*

Kasper, G. (2004). Tax procrastination: Survey finds 29% have yet to begin taxes [Electronic Version] from http://www.prweb.com/releases/2004/03/prweb114250.htm.

Weinstein, G. (2004). *The procrastinator's guide to taxes made easy.* New York: Penguin Group.

14 (2006). *Compound interest, Manhattan & the Indians.* Retrieved from: http://www.savingadvice.com/blog/2006/01/15/10341_compound-interest-manhattan-the-indians.html

15 Byrne, A., Blake, D., Cairns, A., & Dowd, K. (2006). There's no time like the present: The cost of delaying retirement saving *Financial Services Review, 15*(3), 213–231.

16 Lazarus, D. (April 24, 2009). Obama scolds card issuers, and their silence speaks volumes. *Los Angeles Times.* Retrieved from http://www.latimes.com/business/la-fi-lazarus24–2009apr24,0,6516756.column

17 Heidhues, P., & Koszegi, B. (2008). Exploiting naivete about self-control in the credit market. *University of California, Berkeley.*

Shui, H., & Ausubel, L. M. (2005). Time inconsistency in the credit card market. *University of Maryland.*

Spinella, M., Yang, B., & Lester, D. (2004). Prefrontal system dysfunction and credit card debt. *International Journal of Neuroscience, 114,* 1323–1332.

18 Frontline (2008). *The secret history of the credit card.* Retrieved from: http://www.pbs.org/wgbh/pages/frontline/shows/credit/view/

19 Reuben, E., Sapienza, P., & Zingales, L. (2008). Procrastination and impatience: NBER Working Paper.

20 Judson, L. C. (1848). *The moral probe: Or one hundred and two common sense essays on the nature of men and things, interspersed with scraps of science and history.* New York: Published by the author.

21 Matlin, E. (2004). *Procrastinator's guide to wills and estate planning.* New York: Penguin.

22 Like procrastination, this is more common than you think. The American Dental Association indicates that only 12 percent of Americans floss daily and about half don't floss at all.

Harrison, H. C. (2005). *The three-contingency model of self-management.* Unpublished PhD dissertation, Western Michigan University, Kalamazoo, MI.

23 Arce, E., & Santisteban, C. (2006). Impulsivity: A review. *Psicothema, 18*(2), 213–220.

Bickel, W. K., Yi, R., Kowal, B. P., & Gatchalian, K. M. (2008). Cigarette smokers discount past and future rewards symmetrically and more than controls: Is discounting a measure of impulsivity? *Drug and Alcohol Dependence, 96,* 256–262.

Carver, C. S. (2005). Impulse and constraint: Perspectives from personality psychology, convergence with theory in other areas, and potential for integration. *Personality and Social Psychology Review, 9*(4), 312–333.

Chamberlain, S., & Sahakian, B. (2007). The neuropsychiatry of impulsivity. *Current Opinion in Psychiatry, 20*(3), 255.

Enticott, P., & Ogloff, J. (2006). Elucidation of impulsivity. *Australian Psychologist, 41*(1), 3–14.

Schmidt, C. (2003). Impulsivity. In E. F. Coccaro (Ed.), *Aggression: Psychiatric assessment and treatment* (pp. 75–87). New York: Informa Health Care.

Sirois, F. M. (2004). Procrastination and intentions to perform health behaviors: The role of self-efficacy and the consideration of future consequences. *Personality & Individual Differences, 37*(1), 115–128.

Sirois, F. M., & Pychyl, T. A. (2002). *Academic procrastination: Costs to health and well-being.* Paper presented at the American Psychological Association, Chicago.

24 Soble, A. G. (2002). Correcting some misconceptions about St. Augustine's sex life. *Journal of the History of Sexuality, 11*(4), 545–569.

25 Bland, E. (2008). An appraisal of psychological & religious perspectives of self-control. *Journal of Religion and Health, 47*(1), 4–16.

McCullough, M. E., & Willoughby, B. L. B. (2009). Religion, self-regulation, and self-control: Associations, explanations, and implications. *Psychological Bulletin, 135*(1), 69–93.

26 Alternatively, you can open the *Panchatantra* section of the *Mahabharata* to read Vishnu Sharma's words, "The man who is tardy acting where utmost speed is called for, rouses the ire of the gods who would set up obstacles in his way; you can bank on that." Also by Sharma, "Time drinks up the essence of every great and noble action, which ought to be performed, and is delayed in the execution." Gandhi, M. K., Strohmeier, J., & Nagler, M. N. (2000). *The Bhagavad Gita according to Gandhi.* Berkeley, CA: Berkeley Hills Books.

27 Cosan, M. E. (1996). *Ramadhan and Taqwa training.* (H. H. Erkaya, Trans.). Retrieved from http://gumushkhanawidargah.8m.com/books/ramadhan/

28 Similarly, the Islamic scholar Dr. Umar Sulaiman al-Ashqar titles an entire section of his book "Satan hinders the slave from acting by means of procrastination and laziness." As he notes, some of the earliest religious advice underscores the seriousness of procrastination: "Beware of procrastinating. It is the greatest of the soldiers of Satan." Al-Nu'man, A. (2002). *The pillars of Islam.* (A. Fyzeem, Trans., revised, and annotated by I. Poonawala). New Delhi: Oxford University Press. (Original work published 960).

al-Ashqar, U. S. (1998). *World of the Jinn and Devils.* (J. Zarabozo, Trans.). Al-Basheer Publications.

29 Olcott, H. S. (1887). *Golden rules of Buddhism.* London: Theosophical Publishing House.

30 Also, we have Tenzin Gyatso, the fourteenth Dalai Lama: "You must not procrastinate. Rather, you should make preparations so that even if you did die tonight, you would have no regrets." Das, S.

(2000). *Awakening to the sacred: Creating a spiritual life from scratch*. London: Bantam.

31 Giloviqh, T., & Medvec, V. H. (1995). The experience of regret: What, when, and why. *Psychological Review, 102*(2), 379–395.

Roese, N. J., & Summerville, A. (2005). What we regret . . . and why. *Personality and Social Psychology Bulletin, 31*(9), 1273–1285.

32 King, L. A., & Hicks, J. A. (2007). Whatever happened to "What might have been"?: Regrets, Happiness, and Maturity. *American Psychologist, 62*(7), 625–636.

CHAPTER SIX

1 Hayden, A. (2003). International work-time trends: The emerging gap in hours. *Just Labour, 2*, 23–35.

Wasow, B. (2004). Comparing European and U.S. Living Standards (The Century Foundation). Accessed at: http://www.tcf.org/list.asp?type=NC&pubid=596.

2 Malachowski, D. (2005). Wasted time at work costing companies billions. from http://salary.com

3 This is in line with other estimates that also put the cost of procrastination at over $9,000 per employee. D'Abate, C., & Eddy, E. (2007). Engaging in personal business on the job: Extending the presenteeism construct. *Human Resource Development Quarterly, 18*(3), 361–383.

4 Wheelan, C. (2002). *Naked economics: Undressing the dismal science*. New York. W. W. Norton.

5 Critchfield, T., & Kollins, S. (2001). Temporal discounting: Basic research and the analysis of socially important behavior. *Journal of Applied Behavior Analysis, 34*(1), 101–122.

6 Spencer, L. (1955). 10 problems that worry presidents. *Harvard Business Review, 33*, 75–83.

7 Steel, P. & König, C. J. (2006). Integrating theories of motivation. *Academy of Management Review, 31*, 889–913.

8 Lavoie, J. A. A., & Pychyl, T. A. (2001). Cyberslacking and the procrastination superhighway: A web-based survey of online procrastination, attitudes, and emotion. *Social Science Computer Review, 19*(4), 431–444.

Johnson, P. R., & Indvik, J. (2003). The organizational benefits of reducing cyberslacking in the workplace. *Proceedings of the Academy of Organizational Culture, Communications and Conflict,* 7(2), 53–59.

Malachowski, D. (2005). Wasted time at work costing companies billions. from http://salary.com

9 Villano, M. (September 30, 2007). It's only a game, but it's played at work. *New York Times.*

10 Lawler, R. (Monday, June 16, 2008). Cisco sees a zettaflood of IP traffic—driven by video. *Contentinople,* from http://www.contentinople.com/author.asp?section_id=450&doc_id=156555

11 Stelter, B. (January 5, 2008). Noontime web video revitalizes lunch at desk. *New York Times.*

12 Kelly, E. P. (Spring, 2001). Electronic monitoring of employees in the workplace. *National Forum.* Retrieved from: http://findarticles.com/p/articles/mi_qa3651/is_200104/ai_n8939300

13 Ladurantaye, S. (April 2, 2008). Corporate crackdown targets employee surfing: Home e-mail accounts, instant messaging, gaming and video-watching websites . . . they're all on the hit list as employers increasingly restrict what content they permit employees to access. *Globe & Mail.*

14 This corporate "big brother" mentality can become incredibly annoying when you have a legitimate reason to access these sites. My colleague Allen Ponak is a professional labor arbitrator, whose job requires him to mediate a variety of union–management disputes, including when an employee is caught downloading porn to his computer. Part of his job—and I am led to believe he is paid for this—is to examine the content of these sites.

American Management Association (2005). *Electronic monitoring & surveillance survey.* New York: Author.

15 Levin, J. (May 14, 2008). Solitaire-y confinement: Why we can't stop playing a computerized card game. *Slate.*

16 Phillips, J. G., & Reddie, L. (2007). Decisional style and self-reported Email use in the workplace. *Computers in Human Behavior, 23*(5), 2414–2428.

Song, M., Halsey, V., & Burress, T. (2007). *The hamster revolution: How to manage your email before it manages you.* San Francisco: Berrett-Koehler Publishers.

Thatcher, A., Wretschko, G., & Fridjhon, P. (2008). Online flow experiences, problematic Internet use and Internet procrastination. *Computers in Human Behavior, 24,* 2236–2254.

17 Iqbal, S. T., & Horvitz, E. (2007). Conversations amidst computing: A study of interruptions and recovery of task activity. *Proceeds of User Modeling,* 350–354.

18 Richtel, M. (June 14, 2008). Lost in E-mail, tech firms face self-made beast. *New York Times.*

19 Alboher, M. (June 10, 2008). Attention must be paid. *New York Times.*

20 Monsell, S. (2003). Task switching. *TRENDS in Cognitive Sciences,* 7(3), 134–140.

Rubinstein, J. S., Meyer, D. E., & Evans, J. E. (2001). Executive control of cognitive processes in task switching. *Journal of Experimental Psychology: Human Perception and Performance,* 27(4), 763–797.

21 Akerlof, G., & Shiller, R. (2009). *Animal spirits: How human psychology drives the economy, and why it matters for global capitalism.* Princeton, NJ: Princeton University Press.

22 Dunleavy, M. P. (December 2, 2006). Plan to retire but leave out Social Security. *New York Times.*

23 As Avner Offer describes it, "the long-term pattern is that the overall capacity for saving has declined quite substantially since the 1960s, suggesting a declining capacity for prudence."

Offer, A. (2006). *The challenge of affluence: Self-control and well-being in the United States and Britain since 1950.* New York: Oxford University Press.

Weber, E. (2004). Who's afraid of a poor old-age? Risk perception in risk management decisions. In O. Mitchell & S. Utkus (Eds.), *Pension design and structure: New lessons from behavioral finance* (pp. 53–66). New York: Oxford University Press.

24 Transamerica Center for Retirement Studies (2008). The attitudes of American workers and their employers regarding retirement security and benefits. *Ninth Annual Transamerica Retirement Survey.* Available at: http://www.transamericacenter.org/resources/BuildingConfidencePresentation%20TCRS%201002-0208.pdf

25 Brooks, D. (2009). Usury country. *Harper's, 318* (1907), 41–48.

26 Byrne, A., Blake, D., Cairns, A., & Dowd, K. (2006). There's no time like the present: The cost of delaying retirement saving. *Financial Services Review, 15*(3), 213–231.

27 Notably, the economist Matthew Rabin, one of the authors of *Procrastination in Preparing for Retirement,* has included himself among those who aren't saving enough.

O'Donoghue, T., & Rabin, M. (1999). Procrastination in preparing for retirement. In H. J. Aaron (Ed.), *Behavioral dimensions of retirement economics* (pp. 125–156). New York: Brookings Institution Press.

Transamerica Center for Retirement Studies (2008). The attitudes of American workers and their employers regarding retirement security and benefits. *Ninth Annual Transamerica Retirement Survey.* Available at: http://www.transamericacenter.org/resources/BuildingConfidencePresentation%20TCRS%201002-0208.pdf

Organisation of Economic Cooperation and Development (December 2008). Pension Markets in Focus *OECD Newsletter, 5,* 1–20.

28 Byrne, A., Blake, D., Cairns, A., & Dowd, K. (2006). There's no time like the present: the cost of delaying retirement saving. *Financial Services Review, 15*(3), 213–231.

Hewitt Associates (July, 2008). *Hewitt study reveals widening gap between retirement needs and employee saving behaviors.* Retrieved: http://www.businesswire.com/portal/site/google/?ndmViewId=news_view&newsId=20080701005267&newsLang=en

Venti, S. (2006). Choice, Behavior and Retirement Saving. In G. Clark, A. Munnell & M. Orszag (Eds.), *Oxford Handbook of Pensions and Retirement Income* (Vol. 1, pp. 21–30). Oxford: Oxford University Press.

29 O'Donoghue, T., & Rabin, M. (1999). Procrastination in preparing for retirement. In H. J. Aaron (Ed.), *Behavioral dimensions of retirement economics* (pp. 125–156). New York: Brookings Institution Press.

30 Armour, P., & Daly, M. (2008). Retirement savings and decision errors: Lessons from behavioral economics. *FRBSF Economic Letter, 16,* 1–3.

Legorano, G. (2009). Automatic enrollment gains ground for DC plans. *Global Pensions* from http://www.globalpensions.com/global-pensions/news/1557589/automatic-enrollment-gains-ground-dc-plans

Mitchell, O., & Utkus, S. (2003). *Lessons from behavioral finance for retirement plan design.* The Wharton School: University of Pennsylvania.

Turner, J. (2006). Designing 401 (k) plans that encourage retirement savings: Lessons from behavioral finance. *Benefits Quarterly, 22*(4), 1–19.

31 Choi, J., Laibson, D., & Madrian, B. (2004). Plan design and 401 (k) savings outcomes. *National Tax Journal, 57*(2), 275–298.

32 Thaler, R., & Benartzi, S. (2004). Save More Tomorrow™: Using behavioral economics to increase employee saving. *Journal of Political Economy, 112*(S1), 164–187.

33 The debt ceiling was deemed "a meaningless strait jacket" by Robinson as early as 1959.

Austin, D. (2008). *The debt limit: History and recent increases.* Congressional Research Service.

Robinson, M. A. (1959). *The national debt ceiling: An experiment in fiscal policy.* Washington, D. C.: Brookings Institute.

34 Critchfield, T. S., Haley, R., Sabo, B., Colbert, J., & Macropoulis, G. (2003). A half century of scalloping in the work habits of the United States Congress. *Journal of Applied Behavior Analysis, 36,* 465–486.

Weisberg, P., & Waldrop, P. (1972). Fixed-interval work habits of Congress. *Journal of Applied Behavior Analysis, 5*(1), 93–97. Also, special thanks to Tom Critchfield for personally providing me with the data.

35 America's history is particularly steeped in procrastination. During their civil war, procrastination by the General Longstreet cost the South the war when his delays prevented him securing the key positions of Little Round Top and Cemetery Ridge during the Battle of Gettysburg. Meanwhile, Abraham Lincoln struggled with the procrastination of General George Brinton McClellan, which ensured that the war dragged on three extra years. Regarding the procrastination that cost Colonel Rahl his life but in return put America a step closer to independence: there are a few words about this event by British ambassador Nolbert Quayle, "Only a few minutes' delay cost him [Colonel Rahl] his life, his honor, and the liberty of his soldiers. Earth's history is strewn with the wrecks of half-finished plans and unexecuted resolutions. 'Tomorrow' is the excuse of the lazy and refuge of the incompetent." Unfortunately for Quayle, the only record I could find of his existence is this quotation itself.

36 The policy of appeasing Hitler is often characterized as delay that gave the Führer greater time to prepare for battle. Winston Churchill is best known for capturing this sentiment, saying three years prior to Germany's invasion of Poland: "The era of procrastination, of half-measures, of soothing and baffling expedients, of delays, is coming to a close. In its place we are entering a period of consequences . . . We cannot avoid this period; we are in it now." In the aftermath of the war, Dwight D. Eisenhower, former

Supreme Commander of the Allied forces in Europe and thirty-fourth president of the United States, found procrastination still undefeated. The Soviets were preparing themselves for a nuclear confrontation, with little being done to prevent it in Western Europe. Eisenhower's key concern was that the North Atlantic Treaty Organization (NATO) was still just a paper invention, unfunded and militarily toothless. In a speech that Churchill considered to be the greatest he ever heard, at least by an American, we have Eisenhower saying: "The project faces the deadly danger of procrastination, timid measures, slow steps and cautious stages. Granted that the bars of tradition and habit are numerous and stout, the greatest bar to this, or any human enterprise, lie in the minds of men themselves. The negative is always the easy side, since it holds that nothing should be done. The negative is happy in lethargy, contemplating almost with complacent satisfaction, the difficulties of any other course."

37 Andreou, C. (2007). Environmental preservation and second-order procrastination. *Philosophy & Public Affairs, 35*(3), 233–248.

Caney, S. (2008). *Climate change, human rights and intergenerational equity.* Oxford: Magdalen College.

Hepburn, C. (2003). Hyperbolic discounting and resource collapse, *Discussion-Paper No. 159.* Department of Economics, University of Oxford.

Read, D. (2001). Intrapersonal dilemmas. *Human Relations, 54*(8), 1093–1117.

38 Hurni, H., Herweg, K., Portner, B., & Liniger, H. (2008). Soil erosion and conservation in global agriculture. In A. Braimoh & P. L. G. Vlek (Eds.), *Land use and soil resources* (pp. 41–72). New York: Springer.

Montgomery, D. (2007). Soil erosion and agricultural sustainability. *Proceedings of the National Academy of Sciences, 104*(33), 13268–13272.

Sample, I. (August 31, 2007). Global food crisis looms as climate change and population growth strip fertile land. *The Guardian.*

39 Hightower, M. & Pierce, S. A. (2008) The energy challenge. *Nature, 452,* 285–286.

40 Editorial. (March 9, 2008). Oceans at risk. *New York Times.*

Worm, B., Barbier, E., Beaumont, N., Duffy, J., Folke, C., Halpern, B., Jackson, J., Lotze, H., Micheli, F., & Palumbi, S. (2006). Impacts of Biodiversity Loss on Ocean Ecosystem Services. *Science, 314,* 787–790.

Simpson, J. (November 26, 2008). Fishing the fish stocks to extinction. *Globe and Mail.*

41 Lynas, M. (2007). *Six degrees: Our future on a hotter planet.* New York: HarperCollins.

Spratt, D., & Sutton, P. (2008). *Climate Code Red: The case for emergency action.* Melbourne: Scribe Publications.

42 Bamberg, S. (2003). How does environmental concern influence specific environmentally related behaviors? A new answer to an old question. *Journal of Environmental Psychology, 23*(1), 21–32.

Orr, D. W. (2004). *The nature of design: Ecology, culture, and human intention.* New York. Oxford University Press.

43 Farrand, M. (Ed.) (1966). *Records of the federal convention* (Vol. 3). New Haven, CT: Yale University Press.

44 Actually, drinking tea from a saucer only became a social *faux pas* after Washington's and Jefferson's time. Back then, it was quite fashionable to drink tea from a saucer, with accompanying "cup plates" to allow tea drinkers to park their mugs while saucer sipping. Frost, S. (1869). *Frost's laws and by-laws of American society.* New York: Dick & Fitzgerald.

Titus, S. *Tea: A Brief History.* http://www.memorialhall.mass.edu/classroom/curriculum_12th/unit3/lesson8/bkgdessay.html

45 Cumming, L. (2008). *To guide the human puppet: Behavioural economics, public policy and public service contracting:* Serco Institute.

CHAPTER SEVEN

1 Booth, D., & James, R. (2008). A literature review of self-efficacy and effective job search. *Journal of Occupational Psychology, Employment and Disability, 10*(1), 27–42.

Lay, C. H., & Brokenshire, R. (1997). Conscientiousness, procrastination, and person-task characteristics in job searching by unemployed adults. *Current Psychology: Developmental, Learning, Personality, Social, 16*(1), 83–96.

Senecal, C., & Guay, F. (2000). Procrastination in job-seeking: An analysis of motivational processes and feelings of hopelessness. *Journal of Social Behavior & Personality, 15*(5), 267–282.

2 Sigall, H., Kruglanski, A., & Fyock, J. (2000). Wishful thinking and procrastination. *Journal of Social Behavior & Personality, 15*(5), 283–296.

3 Scheier, M. F., & Carver, C. S. (1993). On the power of positive thinking: The benefits of being optimistic. *Current Directions in Psychological Science, 2*(1), 26–30.

4 Despite the fact that we all tend to underestimate the time it takes, procrastinators tend to be worse at this.

Buehler, R., Griffin, D., & Ross, M. (1994). Exploring the "planning fallacy": Why people underestimate their task completion times. *Journal of Personality and Social Psychology, 67*, 366–381.

Kahneman, D., & Tversky, A. (1979). Intuitive prediction: Biases and corrective procedures. *TIMS Studies in Management Sciences, 12*, 313–327.

Lay, C. H., & Schouwenburg, H. C. (1993). Trait procrastination, time management, and academic behavior. *Journal of Social Behavior & Personality, 8*(4), 647–662.

Roy, M. M., Christenfeld, N. J. S., & McKenzie, C. R. M. (2005). Underestimating the duration of future events: Memory incorrectly used or memory bias? *Psychological Bulletin, 131*(5), 738–756.

Sigall, H., Kruglanski, A., & Fyock, J. (2000). Wishful thinking and procrastination. *Journal of Social Behavior & Personality, 15*(5), 283–296.

5 Vancouver, J., More, K., & Yoder, R. (2008). Self-efficacy and resource allocation: Support for a nonmonotonic, discontinuous model. *Journal of Applied Psychology, 93*(1), 35–47.

6 Ehrlinger, J., Johnson, K., Banner, M., Dunning, D., & Kruger, J. (2008). Why the unskilled are unaware: Further explorations of (absent) self-insight among the incompetent. *Organizational Behavior and Human Decision Processes, 105*(1), 98–121.

Kruger, J., & Dunning, D. (1999). Unskilled and unaware of it: How difficulties in recognizing one's own incompetence lead to inflated self-assessments. *Journal of Personality and Social Psychology, 77*(6), 1121–1134.

Russell, B., & Branch, T. (1991). *Second wind: The memoirs of an opinionated man.* New York: Simon & Schuster.

7 Wegner, D. M. (1994). *White bears and other unwanted thoughts: Suppression, obsession, and the psychology of mental control.* New York: Guilford Press.

Wood, J. V., Perunovic, W. Q. E., & Lee, J. W. (2009). Positive self-statements: Power for some, peril for others. *Psychological Science, 20*(7), 860–866.

Most poignantly illustrative of the potential danger of aphorisms is the story of Felix Powell, a British Staff Sergeant who wrote the music of the morale-building marching song "Pack Up Your Troubles in Your Old Kit Bag and Smile, Smile, Smile," one of the most optimistic songs ever written. Dressed in the uniform of the Peacehaven Home Guard, Powell shot himself in the heart with a rifle, committing suicide. Indeed, positive self-statements can make matters worse for those with low self-esteem.

8 In business academia, they call success spirals an "efficacy-performance deviation amplifying loop."

Lindsley, D., Brass, D. J., & Thomas, J. B. (1995). Efficacy-performance spirals: A multilevel perspective. *Academy of Management Review, 20*(3), 645–678.

9 For early stages of a complex venture, it is often best to have process or learning goals rather than product or outcome goals. That is, the goals are acquiring or refining new skills or steps (the process) rather than winning or getting the highest score (the product). Not only will confidence be maximized but, in the end, higher performance results.

Schunk, D., & Meece, J. (2006). Self-efficacy development in adolescences. In F. Pajares & T. Urdan (Eds.), *Self-efficacy beliefs of adolescents* (pp. 71–96). Greenwich CT: Information Age.

Seijts, G. H. (2001). Setting goals when performance doesn't matter. *Ivey Business Journal, 65*(3), 40–47.

10 Hans, T. A. (2000). A meta-analysis of the effects of adventure programming on locus of control. *Journal of Contemporary Psychotherapy, 30*(1), 33–60.

Hattie, J., Marsh, H. W., Neil, J. T., & Richards, G. E. (1997). Adventure education and Outward Bound: Out-of-class experiences that make a lasting difference. *Review of Educational Research, 67*(1), 43–87.

Wilson, S. J., & Lipsey, M. W. (2000). Wilderness challenge programs for delinquent youth: A meta-analysis of outcome evaluations. *Evaluation and Program Planning, 23,* 1–12.

11 Feldman, A., & Matjasko, J. (2005). The role of school-based extracurricular activities in adolescent development: A comprehensive review and future directions. *Review of Educational Research, 75*(2), 159–210.

12 World Organization of the Scout Movement (1998). *Scouting: An educational system.* Geneva, Switzerland: World Scout Bureau.

13 Gestdottir, S., & Lemer, R. M. (2007). Intentional self-regulation and positive youth development in early adolescence: Findings from the 4-H study of positive youth development. *Developmental Psychology, 43*(2), 508–521.

Jelicic, H., Bobek, D., Phelps, E., Lerner, R., & Lerner, J. (2007). Using positive youth development to predict contribution and risk behaviors in early adolescence: Findings from the first two waves of the 4-H Study of Positive Youth Development. *International Journal of Behavioral Development, 31*(3), 263–273.

Radhakrishna, R., & Sinasky, M. (2005). 4-H experiences contributing to leadership and personal development of 4-H alumni. *Journal of Extension, 43*(6). Retrieved from: http://www.joe.org/joe/2005december/rb2.php

14 Zimmerman, B. J. (2002). Becoming a self-regulated learner: An overview. *Theory into Practice, 41*(2), 64–70.

15 Early efforts to combat procrastination often focused on just this one step, using cognitive therapy to challenge people's self-limiting beliefs. It was notably used by the late Albert Ellis, whose approach is being continued by his co-author, William Knaus.

Ellis, A., & Knaus, W. J. (1977). *Overcoming procrastination: Or how to think and act rationally in spite of life's inevitable hassles.* Institute for Rational Living.

16 Schunk, D., & Meece, J. (2006). Self-efficacy development in adolescences. In F. Pajares & T. Urdan (Eds.), *Self-efficacy beliefs of adolescents* (pp. 71–96). Greenwich CT: Information Age.

17 This includes the leaders we follow as much as the spouses we choose (e.g., "Behind every great man/woman, there is a great woman/man"). Aside from role models and comparison groups being key determinants of self-efficacy, what others believe (that is, normative beliefs and subjective norms) play a major role in forming an intention to act.

Aarts, H., Dijksterhuis, A., & Dik, G. (2008). Goal contagion: Inferring goals from others' actions—and what it leads to. In J. Y. Shah & W. L. Gardner (Eds.), *Handbook of motivation* (pp. 265–280). New York: Guilford Press.

Armitage, C., & Conner, M. (2001). Efficacy of the theory of planned behaviour: A meta-analytic review. *British Journal of Social Psychology, 40*(4), 471–499.

Rivis, A., & Sheeran, P. (2003). Descriptive norms as an additional predictor in the theory of planned behaviour: A meta-analysis. *Current Psychology, 22*(3), 218–233.

van Knippenberg, D., van Knippenberg, B., De Cremer, D., & Hogg, M. (2004). Leadership, self, and identity: A review and research agenda. *The Leadership Quarterly, 15*(6), 825–856.

18 Vitale, J., & Hibbler, B. (2006). *Meet and grow rich: How to easily create and operate your own "Mastermind" group for health, wealth, and more.* Hoboken, NJ: John Wiley & Sons.

19 Metta, G., Sandini, G., Natale, L., Craighero, L., & Fadiga, L. (2006). Understanding mirror neurons. *Interaction Studies, 7*(2), 97–232.

Weinberg, R. (2008). Does imagery work? Effects on performance and mental skills. *Journal of Imagery Research in Sport and Physical Activity, 3*(1), 1–21.

20 Achtziger, A., Fehr, T., Oettingen, G., M. Gollwitzer, P., & Rockstroh, B. (2008). Strategies of intention formation are reflected in continuous MEG activity. *Social Neuroscience, 4(*1), 1–17.

Oettingen, G., Mayer, D., Thorpe, J. S., Janetzke, H., & Lorenz, S. (2005). Turning fantasies about positive and negative futures into self-improvement goals. *Motivation and Emotion, 29*(4), 236–266.

Oettingen, G., & Thorpe, J. S. (2006). Fantasy realization and the bridging of time. In L. A. Sanna, & E. C. Chang (Eds.), *Judgments over time: The interplay of thoughts, feelings, and behaviors* (pp. 120–143). Oxford: Oxford University Press. But also: Kavanagh, D. J., Andrade, J., & May, J. (2005). Imaginary relish and exquisite torture: The elaborated intrusion theory of desire. *Psychological Review, 112*(2), 446–467.

Pham, L. B., & Taylor, S. E. (1999). From thought to action: Effects of process- versus outcome-based mental simulations on performance. *Personality and Social Psychology Bulletin, 25*, 250–260.

21 It may also be a bad idea to promote a pattern of thinking that puts a person at increased risk for a wide variety of mental illnesses. In compensation, though, those who are extremely fantasy-prone can enjoy imagined food as much as the real and can imagine themselves to orgasm without physical stimulation.

Levin, R., & Spei, E. (2004). Relationship of purported measures of pathological and nonpathological dissociation to self-reported psychological distress and fantasy immersion. *Assessment, 11*(2), 160–168.

Rhue, J., & Lynn, S. (1987). Fantasy proneness: The ability to hallucinate "as real as real." *British Journal of Experimental and Clinical Hypnosis, 4,* 173–180.

Schneider, S. L. (2001). In search of realistic optimism. Meaning, knowledge, and warm fuzziness. *American Psychologist, 56*(3), 250–263.

Waldo, T. G., & Merritt, R. D. (2000). Fantasy proneness, dissociation, and DSM-IV axis II symptomatology. *Journal of Abnormal Psychology, 109*(3), 555–558.

22 Johnson, D. D. P. (2004). *Overconfidence and war: The havoc and glory of positive illusions.* Cambridge, MA: Harvard University Press.

23 Armor, D., & Taylor, S. (2002). When predictions fail: The dilemma of unrealistic optimism. In T. Gilovich, D. Griffin & D. Kahneman (Eds.), *Heuristics and biases: The psychology of intuitive judgment* (pp. 334–347). New York: Cambridge University Press.

Asterbro, T., Jeffrey, S., & Adomdza, G. K. (2007). Inventor perseverance after being told to quit: The role of cognitive biases. *Journal of Behavioral Decision Making, 20*(3), 253–272.

Lovallo, D., & Kahneman, D. (2003). Delusions of success. How optimism undermines executives' decisions. *Harvard Business Review, 81*(7), 56–63.

Moore, D., & Healy, P. (2007). *The trouble with overconfidence.* Unpublished manuscript, Carnegie-Mellon University, Pittsburgh.

24 Baker, W., & O'Malley, M. (2008). *Leading with kindness: How good people consistently get superior results.* New York: AMACOM/American Management Association.

Whyte, G., Saks, A., & Hook, S. (1997). When success breeds failure: The role of self-efficacy in escalating commitment to a losing course of action. *Journal of Organizational Behavior, 18*(5), 415–432.

25 Camerer, C. F., & Lovallo, D. (1999). Overconfidence and excess entry: An experimental approach. *American Economic Review, 89*(1), 306–318.

Koellinger, P., Minniti, M., & Schade, C. (2007). "I think I can, I think I can": Overconfidence and entrepreneurial behavior. *Journal of Economic Psychology, 28*(4), 502–527.

Hmieleski, K., & Baron, R. (2009). Entrepreneurs' optimism and new venture performance: A social cognitive perspective. *Academy of Management Journal, 52*(3), 473–488.

Shepherd, D. A., Wiklund, J., & Haynie, J. M. (2009). Moving forward: Balancing the financial and emotional costs of business failure. *Journal of Business Venturing, 24*(2), 134–148.

26 Day, V., Mensink, D., & O'Sullivan, M. (2000). Patterns of academic procrastination. *Journal of College Reading and Learning, 30*(2), 120–134.

Sigall, H., Kruglanski, A., & Fyock, J. (2000). Wishful thinking and procrastination. *Journal of Social Behavior & Personality, 15*(5), 283–296.

27 Though having several critics, such as the influential psychologist Albert Ellis, and accused of being a confidence (con) man, Peale's popularity still remains strong.

Hilkey, J. (1997). *Character is capital: Success manuals and manhood in Gilded Age America.* Chapel Hill: University of North Carolina Press.

Meyer, D. (1988). *The positive thinkers: Popular religious psychology from Mary Baker Eddy to Norman Vincent Peale and Ronald Reagan.* Middletown, CT: Wesleyan University Press.

Weiss, R. (1988). *The American myth of success: From Horatio Alger to Norman Vincent Peale.* Urbana, IL: University of Illinois Press.

28 Barbara Held, a psychology professor at Bowdoin College, describes it this way: "The positive attitude has—in some of its manifestations—become tyrannical, in that Americans have come to live not only with a historically/culturally grounded inclination for optimism but with the expectation, with the demand, that they maintain a positive attitude at all times and at all costs."

De Raeve, L. (1997). Positive thinking and moral oppression in cancer care. *European Journal of Cancer Care, 6*(4), 249–256.

Ehrenreich, B. (2009). *Bright-sided: How the relentless promotion of positive thinking has undermined America.* New York: Metropolitan Books.

Fineman, S. (2006). On being positive: Concerns and counterpoints. *The Academy of Management Review, 31*(2), 270–291.

Gilovich, T. (2005). *The perceived likelihood of events that "tempt fate."* Paper presented at the Annual Meeting of the Society of Personality and Social Psychology, New Orleans.

Held, B. (2002). The tyranny of the positive attitude in America: Observation and speculation. *Journal of Clinical Psychology, 58*(9), 965–991.

Recken, S. L. (1993). Fitting-in: The redefinition of success in the 1930s. *Journal of Popular Culture, 27*(3), 205–222.

Woolfolk, R. L. (2002). The power of negative thinking: Truth, melancholia, and the tragic sense of life. *Journal of Theoretical and Philosophical Psychology, 22*(1), 19–27.

29 Nenkov, G. Y., Inman, J. J., & Hulland, J. (2008). Considering the future: The conceptualization and measurement of elaboration on potential outcomes. *Journal of Consumer Research, 35*(1), 126–141.

Pearson, C. M., & Clair, J. A. (1998). Reframing crisis management. *The Academy of Management Review, 23*(1), 59–76.

Schneider, S. L. (2001). In search of realistic optimism. Meaning, knowledge, and warm fuzziness. *American Psychologist, 56*(3), 250–263.

Yordanova, G. S. (2006). *Effects of the pre-decision stage of decision making on the self-regulation of behavior.* Unpublished PhD, University of Pittsburgh, Pittsburgh, PN.

30 Jones, F., Harris, P., Waller, H., & Coggins, A. (2005). Adherence to an exercise prescription scheme: The role of expectations, self-efficacy, stage of change and psychological well-being. *British Journal of Health Psychology, 10,* 359–378.

Nordgren, L. F., Harreveld, F. V., & Pligt, J. V. D. (2009). The restraint bias: How the illusion of self-restraint promotes impulsive behavior. *Psychological Science, 20,* 1523–1528.

Norcross, J. C., Mrykalo, M. S., & Blagys, M. D. (2002). *Auld Lang Syne:* Success predictors, change processes, and self-reported outcomes of New Year's resolvers and nonresolvers. *Journal of Clinical Psychology, 58*(4), 397–405.

Norcross, J. C., Ratzin, A. C., & Payne, D. (1989). Brief report ringing in the New Year: The change processes and reported outcomes of resolutions. *Addictive Behaviors, 14,* 205–212.

Polivy, J., & Herman, C. P. (2002). If at first you don't succeed: False hopes of self-change. *American Psychologist, 57*(9), 677–689.

31 Aspinwall, L. G. (2005). The psychology of future-oriented thinking: From achievement to proactive coping, adaptation, and aging. *Motivation and Emotion, 29*(4), 203–235.

Aspinwall, L. G., & Taylor, S. E. (1997). A stitch in time: Self-regulation and proactive coping. *Psychological Bulletin, 121,* 417–436.

Baumeister, R. F., Heatherton, T. F., & Tice, D. M. (1994). *Losing control: How and why people fail at self-regulation.* San Diego, CA: Academic Press, Inc.

Klassen, R. M., Krawchuk, L. L., & Rajani, S. (2008). Academic procrastination of undergraduates: Low self-efficacy to self-regulate predicts higher levels of procrastination. *Contemporary Educational Psychology, 33*(4), 915–931.

Schwarzer, R. (2008). Modeling health behavior change: How to predict and modify the adoption and maintenance of health behaviors. *Applied Psychology: An International Review, 57*(1), 1–29.

32 Known as the *abstinence violation effect.*

Larimer, M. E., Palmer, R. S., & Marlatt, G. A. (1999). Relapse prevention: An overview of Marlatt's cognitive-behavioral model. *Alcohol Research & Health, 23*(2), 151–160.

33 Howard Rachlin gives a similar account under the rubric of "restructuring" and Jeong-Yoo Kim considers the same phenomenon from an economic perspective. Another pair of economists, Benabou and Tirole, discuss how it is best to assume that you don't have the self-control to resist possible addictions, even if there is a good chance you could use without risk. Interestingly, Buddhists actually use an enhanced form of this technique by believing bad choices (i.e., karma) will not only negatively impact your future self but also your future reincarnations.

Ainslie, G. (1992). *Picoeconomics: The strategic interaction of successive motivational states within the person.* New York: Cambridge University Press.

Ainslie, G. (2001). *Breakdown of the will.* New York: Cambridge University Press.

Benabou, R., & Tirole, J. (2004). Willpower and personal rules. *Journal of Political Economy, 112*(4), 848–886.

Kim, J.-Y. (2006). Hyperbolic discounting and the repeated self-control problem. *Journal of Economic Psychology, 27*(3), 344–359.

Rachlin, H. (2000). *The science of self-control.* Cambridge, MA: Harvard University Press.

34 Gosling, J. (1990). *Weakness of the will.* New York: Routledge.

35 Silver, M., & Sabini, J. (1981). Procrastinating. *Journal for the Theory of Social Behavior, 11*(2), 207–221.

CHAPTER EIGHT

1 Fried, Y., & Ferris, G. R. (1987). The validity of the Job Characteristics Model: A review and meta-analysis. *Personnel Psychology, 40*(2), 287–322.

Hackman, J. R., & Oldham, G. R. (1976). Motivation through the design of work: Test of a theory. *Organizational Behavior and Human Performance, 16*, 250–279.

Humphrey, S., Nahrgang, J., & Morgeson, F. (2007). Integrating motivational, social, and contextual work design features: A meta-analytic summary and theoretical extension of the work design literature. *Journal of Applied Psychology, 92*(5), 1332–1356.

2 Others were involved, such as Frank and Lillian Gilbreth who pioneered time and motion studies. The Gilbreths' work and life were chronicled in a book, *Cheaper by the Dozen,* written by two of their twelve children (Frank Jr. and Ernestine). Lillian was arguably the first of my kind—an Industrial/Organizational Psychologist—getting a PhD in management psychology (as well as receiving twenty-two other honorary degrees). The book became a film in 1950, not to be confused with the 2003 feature by the same name. This later version, starring Steve Martin and Bonnie Hunt, has some changes. Instead of Industrial/Organizational Psychology, this adaptation centers around a football coach because apparently there just aren't enough movies produced each year featuring football.

Kanigel, R. (1997). *The one best way: Frederick Winslow Taylor and the enigma of efficiency.* New York: Viking Penguin.

3 Furthermore, the harder employees worked, the less they were paid for each unit they produced. This is the typical outcome of most piece-rate systems, whereby you get paid for what your produce. Paradoxically, it is an inherent temptation for managers to reduce incentives as employees provide the very performance they were trying to incent. Known as the *rachet effect,* only a very few companies, like Lincoln Electric, have the discipline to avoid it and make the piece-rate system work.

Handlin, H. (1992). The company built upon the golden rule: Lincoln Electric. *Journal of Organizational Behavior Management, 12*, 151–163.

Billikopf, G. (2008). *Designing an effective piece rate.* Retrieved from: http://www.cnr.berkeley.edu/ucce50/ag-labor/7research/7calag06.htm

4 Campion, M., Mumford, T., Morgeson, F., & Nahrgang, J. (2005). Work redesign: Eight obstacles and opportunities. *Human Resource Management, 44*(4), 367–390.

5 Cosmides, L., & Tooby, J. (2000). Evolutionary psychology and the emotions. In M. Lewis & J. Haviland (Eds.), *Handbook of Emotions* (2 ed., pp. 91–115). New York: Guilford Press.

6 Science studies the malleable nature of value under the term "psychophysics," with research emphasizing, as here, that value is constructed (that is, dependent on how it is presented) and relative (i.e., dependent on what it is being compared to).

Weber, E. (2003). Perception matters: Psychophysics for economists. In I. Brocas & J. D. Carrillo (Eds.), *The Psychology of Economic Decisions* (Vol. II). New York: Oxford University Press.

7 Sansone, C., Weir, C., Harpster, L., & Morgan, C. (1992). Once a boring task always a boring task? Interest as a self-regulatory mechanism? *Journal of Personality & Social Psychology, 63*(3), 379–390.

8 Csíkszentmihályi, M. (1990). *Flow: The psychology of optimal experience.* New York: Harper and Row.

9 Johnny Carson of "The Tonight Show" invited her as a guest and pretended to eat her prized Elvis Presley chip. CNN (January 24, 2005). Your Johnny Carson memories. Retrieved from: http://www.cnn.com/2005/SHOWBIZ/TV/01/23/your.memories/index.html

10 Miller, R. B., & Brickman, S. J. (2004). A model of future-oriented motivation and self-regulation. *Educational Psychology Review, 16*(1), 9–33.

Schraw, G., & Lehman, S. (2001). Situational interest: A review of the literature and directions for future research. *Educational Psychology Review, 13*(1), 23–52.

Wolters, C. A. (2003). Understanding procrastination from a self-regulated learning perspective. *Journal of Educational Psychology, 95*(1), 179–187.

11 Ryan, R. M., & Deci, E. L. (2000). Self-determination theory and the facilitation of intrinsic motivation, social development, and well-being. *American Psychologist, 55*(1), 68–78.

12 Lonergan, J. M., & Maher, K. J. (2000). The relationship between job characteristics and workplace procrastination as moderated by locus of control. *Journal of Social Behavior & Personality, 15*(5), 213–224.

Miller, R. B., & Brickman, S. J. (2004). A model of future-oriented motivation and self-regulation. *Educational Psychology Review, 16*(1), 9–33.

Shah, J., & Kruglanski, A. (2000). The structure and substance of intrinsic motivation. In C. Sansone & J. M. Harackiewicz (Eds.), *Intrinsic and extrinsic motivation: The search for optimal motivation and performance* (pp. 106–130). San Diego, CA: Academic Press.

13 I like Franklin Jones' quote to this effect: "Nothing makes it easier to resist temptation than a proper bringing-up, a sound set of values—and witnesses."

Becker, H. (1960). Notes on the concept of commitment. *American Journal of Sociology, 66*(1), 32–40.

Magen, E., & Gross, J. J. (2007). Harnessing the need for immediate gratification: Cognitive reconstrual modulates the reward value of temptations. *Emotion, 7*(2), 415–428.

Powell, D., & Meyer, J. (2004). Side-bet theory and the three-component model of organizational commitment. *Journal of Vocational Behavior, 65*(1), 157–177.

14 Newman, T. (December 20, 2008). Barack Obama, I quit smoking—all the time. *Newsday.* Retrieved from http://www.newsday.com/news/opinion/ny-opnew205971623dec20,0,6796122.story.

15 Elliot, A., & Friedman, R. (2006). Approach-avoidance: A central characteristic of personal goals. In B. R. Little, K. Salmela-Aro & S. D. Phillips (Eds.), *Personal project pursuit: Goals, action, and human flourishing* (pp. 97–118). Mahwah, NJ: Lawrence Erlbaum Associates.

Howell, A. J., & Watson, D. C. (2007). Procrastination: Associations with achievement goal orientation and learning strategies. *Personality and Individual Differences, 43*(1), 167–178.

Mogilner, C., Aaker, J., & Pennington, G. (2007). Time will tell: The distant appeal of promotion and imminent appeal of prevention. *Journal of Consumer Research, 34*(5), 670–681.

Polivy, J., & Herman, C. P. (2002). If at first you don't succeed: False hopes of self-change. *American Psychologist, 57*(9), 677–689.

Schneider, S. L. (2001). In search of realistic optimism. Meaning, knowledge, and warm fuzziness. *American Psychologist, 56*(3), 250–263.

Wolters, C. A. (2003). Understanding procrastination from a self-regulated learning perspective. *Journal of Educational Psychology, 95*(1), 179–187.

Wolters, C. A. (2004). Advancing achievement goal theory: Using goal structures and goal orientations to predict students' motivation, cognition, and achievement. *Journal of Educational Psychology, 96*(2), 236–250.

Valkyrie, K. T. (2006). *Self-regulated learning: An examination of motivational, cognitive, resource management, metacognitive components and academic outcomes with open admissions community college students.* Unpublished PhD dissertation, University of Houston, Houston, TX.

16 Also, you can further upgrade your approach goals by making them about mastery. Mastery is viewing life as a prolonged opportunity to improve, to live to your potential. Each challenge, won or lost, is another step toward consummate skill. Mastery goals much more reliably produce the intrinsic motivation you are looking for. Similarly, those who are already at the top can eke out a

little extra motivation by framing their approach goals in terms of prevention; that is, achievement will *prevent* them from losing their desirable position. Goals that emphasize protecting and maintaining standing and success will help you start a little earlier than everyone else.

Freitas, A. L., Liberman, N., Salovey, P., & Higgins, E. T. (2002). When to begin? Regulatory focus and initiating goal pursuit. *Personality and Social Psychology Bulletin, 28*(1), 121–130.

Molden, D. C., Lee, A. Y., & Higgins, E. T. (2007). Motivations for promotion and prevention. In W. L. G. James Y. Shah (Ed.), *Handbook of motivation science* (pp. 169–187). New York: Guilford Press.

Rawsthorne, L., & Elliot, A. (1999). Achievement goals and intrinsic motivation: A meta-analytic review. *Personality and Social Psychology Review, 3*(4), 326–344.

Pennington, G. L., & Roese, N. J. (2003). Regulatory focus and temporal distance. *Journal of Experimental Social Psychology, 39*, 563–576.

17 Steel, P. (2007). The nature of procrastination: A meta-analytic and theoretical review of quintessential self-regulatory failure. *Psychological Bulletin, 133*(1), 65–94.

18 Gröpel, P., & Steel, P. (2008). A mega-trial investigation of goal setting, interest enhancement, and energy on procrastination. *Personality and Individual Differences, 45*, 406–411.

19 Reduced energy is another reason why, aside from reduced self-confidence as per the last chapter, depression is connected to procrastination.

Thase, M. E. (1995). Cognitive behavior therapy. In I. D. Glick (Ed.), *Treating depression* (pp. 33–70). San Francisco: Jossey-Bass, Inc.

20 A *Kids in the Hall* comedy sketch called “Chocolate” depicts this back and forth between wanting to diet and to wanting to eat chocolate. After a few bites, our protagonist throws away his chocolate bar, only to change his mind over and over.

21 Ramanathan, S., & Menon, G. (2006). Time-varying effects of chronic hedonic goals on impulsive behavior. *Journal of Marketing Research, 43*(4), 628–641.

22 Furnham, A. (2002). *Personality at work: The role of individual differences in the workplace.* New York: Routledge.

23 Díaz-Morales, J., Ferrari, J., & Cohen, J. (2008). Indecision and avoidant procrastination: The role of morningness-eveningness and time perspective in chronic delay lifestyles. *Journal of General Psychology, 135*(3), 228–240.

Digdon, N., & Howell, A. (2008). College students who have an eveningness preference report lower self-control and greater procrastination. *Chronobiology International, 25*(6), 1029–1046.

Ferrari, J. R., Harriott, J. S., Evans, L., Lecik-Michna, D. M., & Wenger, J. M. (1997). Exploring the time preferences of procrastinators: Night or day, which is the one? *European Journal of Personality, 11*(3), 187–196.

Hess, B., Sherman, M. F., & Goodman, M. (2000). Eveningness predicts academic procrastination: The mediating role of neuroticism. *Journal of Social Behavior and Personality, 15*(5), 61–74.

24 Klein, S. (2009). *The secret pulse of time: Making sense of life's scarcest commodity.* Cambridge, MA: Da Capo Lifelong Books.

25 Oaten, M., & Cheng, K. (2006). Longitudinal gains in self-regulation from regular physical exercise. *British Journal of Health Psychology, 11*(4), 717–733.

26 Though Jim Horne, from the University of Loughborough's Sleep Research Centre, contends we are actually sleeping better now than in most of history.

Horne, J. (18 October, 2008). Time to wake up to the facts about sleep. *New Scientist, 2678*, 36–38.

Mooallem, J. (November 18, 2007). The sleep-industrial complex. *The New York Times.*

National Sleep Foundation (2008). Sleep in America Poll. Retrieved from http://www.sleepfoundation.org/atf/cf/%7Bf6bf2668-a1b4-4fe8-8d1a-a5d39340d9cb%7D/2008%20POLL%20SOF.PDF

27 Muris, P., Merckelbach, H., Ollendick, T., King, N., & Bogie, N. (2001). Children's nighttime fears: Parent-child ratings of frequency, content, origins, coping behaviors and severity. *Behaviour Research and Therapy, 39*(1), 13–28.

Tooby, J., & Cosmides, L. (1990). The past explains the present: Emotional adaptations and the structure of ancestral environments. *Ethology and Sociobiology, 11*(4–5), 375–424.

28 Bettelheim, B. (1977). *The uses of enchantment: The meaning and importance of fairy tales.* New York: Knopf.

29 Ferrari, J. R., & McCown, W. (1994). Procrastination tendencies among obsessive-compulsives and their relatives. *Journal of Clinical Psychology, 50*(2), 162–167.

Rachman, S. (1993). Obsessions, responsibility and guilt. *Behaviour Research & Therapy, 31*(2), 149–154.

Kaplan, A., & Hollander, E. (2004). Comorbidity in compulsive hoarding: a case report. *CNS Spectrums, 9*(1), 71–73.

30 Benton, T. H. (2005). Productive procrastination. *The Chronicle of Higher Education, 52*(1).

31 Bandura, A. (1976). Self-reinforcement: Theoretical and methodological considerations. *Behaviorism, 4*(2), 135–155.

Febbraro, G., & Clum, G. (1998). Meta-analytic investigation of the effectiveness of self-regulatory components in the treatment of adult problem behaviors. *Clinical Psychology Review, 18*(2), 143–161.

Ferrari, J. R., & Emmons, R. A. (1995). Methods of procrastination and their relation to self-control and self-reinforcement: An exploratory study. *Journal of Social Behavior & Personality, 10*(1), 135–142.

32 Eisenberger, R. (1992). Learned industriousness. *Psychological Review, 99*, 248–267.

Renninger, K. (2000). Individual interest and its implications for understanding intrinsic motivation. In C. Sansone & J. M. Harackiewicz (Eds.), *Intrinsic and extrinsic motivation: The search for optimal motivation and performance* (pp. 373–404). San Diego, CA: Academic Press.

Stromer, R., McComas, J. J., & Rehfeldt, R. A. (2000). Designing interventions that include delayed reinforcement: Implications of recent laboratory research. *Journal of Applied Behavior Analysis, 33*, 359–371.

33 Technically known as impulse pairing or fusion.

Ainslie, G. (1992). *Picoeconomics: The strategic interaction of successive motivational states within the person.* New York: Cambridge University Press.

Murray, H. A. (1938). *Explorations in personality.* New York: Oxford University Press.

34 Though this is the common term, some practitioners find it derogatory, dismissing the skill it can involve.

35 Dibbell, J. (June 17, 2007). The life of the Chinese gold farmer. *The New York Times Magazine.*

Jin, G. (2006). Chinese gold farmers in the game world [Electronic Version]. *Consumers, Commodities & Consumption* 7. Retrieved from https://netfiles.uiuc.edu/dtcook/www/CCCnewsletter/7–2/jin.htm.

Jin, G. (2008). *Gold farmers.* Retrieved from http://chinesegoldfarmers.com/Index.html

36 Akerman, D. S., & Gross, B. L. (2007). I can start that JME manuscript next week, can't I? The task characteristics behind why faculty procrastinate. *Journal of Marketing Education, 29*(2), 97–110.

Sansone, C., & Harackiewicz, J. (2000). *Intrinsic and extrinsic motivation: The search for optimal motivation and performance.* San Diego, CA: Academic Press.

37 Bordens, K., & Horowitz, I. (2001). *Social psychology.* Mahwah, NJ: Lawrence Erlbaum Associates.

Moreland, R. L., & Beach, S. R. (1992). Exposure effects in the classroom: The development of affinity among students. *Journal of Experimental Social Psychology, 28*(3), 255–276.

38 Fouad, N. (2007). Work and vocational psychology: Theory, research, and applications. *Annual Review of Psychology, 58,* 543–564.

39 If you want to know your own profile, there are a variety of free versions available online. Just conduct an Internet search using the term "RIASEC."

40 Lubinski, D., & Benbow, C. P. (2000). States of excellence. *American Psychologist, 55*(1), 137–150.

41 It is possible to select a job for you with considerably more accuracy than currently available, directing you to areas of work that you would love as well as those in which you would excel. Unfortunately, despite being designed, proven, and patented, such a system has yet to be built. Sorry for this, but I have been busy writing a book. The patent number is US 20080027771. Interested parties should contact University Technologies International (tech@uti.ca).

Scherbaum, C. A. (2005). Synthetic validity: Past, present, and future. *Personnel Psychology, 58*(2), 481–515.

Steel, P. D., Huffcutt, A. I., & Kammeyer-Mueller, J. (2006). From the work one knows the worker: A systematic review of the challenges, solutions, and steps to creating synthetic validity. *International Journal of Selection and Assessment, 14*(1), 16–36.

Steel, P., & Kammeyer-Mueller, J. (2009). Using a meta-analytic perspective to enhance Job Component Validation. *Personnel Psychology, 62*(3), 533–552.

42 Tullier, L. (2000). *The complete idiot's guide to overcoming procrastination.* Indianapolis, IN: Alpha Books.

CHAPTER NINE

1 Akerlof, G. A. (1991). Procrastination and obedience. *American Economic Review, 81,* 1–19.

Arneklev, B., Elis, L., & Medlicott, S. (2006). Testing the General Theory of Crime: Comparing the effects of "imprudent behavior" and an attitudinal indicator of "low self-control." *Western Criminology Review, 7*(3), 41–55.

Carver, C. S. (2005). Impulse and constraint: Perspectives from personality psychology, convergence with theory in other areas, and potential for integration. *Personality and Social Psychology Review, 9*(4), 312–333.

Glomb, T., Steel, P., & Arvey, R. (2002). Office sneers, snipes, and stab wounds: Antecedents, consequences, and implications of workplace violence and aggression. In R. G. Lord, R. J. Klimoski, & R. Kanfer (Eds.), *Emotions in the workplace: Understanding the structure and role of emotions in organizational behavior* (pp. 227–259). San Francisco, CA: Jossey-Bass.

Gottfredson, M. R., & Hirschi, T. (1990). *A General Theory of Crime.* Stanford, CA: Stanford University Press.

Hirschi, T. (2004). Self-control and crime. In R. F. Baumeister & K. D. Vohs (Eds.), *Handbook of self-regulation: Research, theory, and applications* (pp. 537–552). New York: Guilford Press.

Schmidt, C. (2003). Impulsivity. In E. F. Coccaro (Ed.), *Aggression: Psychiatric assessment and treatment* (pp. 75–87). New York: Informa Health Care.

2 Roberts, B. W., Walton, K. E., & Viechtbauer, W. (2006). Patterns of mean-level change in personality traits across the life course: A meta-analysis of longitudinal studies. *Psychological Bulletin, 132,* 1–25.

3 Funder, D. C. (2001). Personality. *Annual Review of Psychology, 52,* 197–221.

4 Ainslie, G. (1975). Specious reward: A behavioral theory of impulsiveness and impulse control. *Psychological Bulletin, 82*(4), 463–496.

5 Ariely, D., & Wertenbroch, K. (2002). Procrastination, deadlines, and performance: Self-control by precommitment. *Psychological Science, 13*(3), 219–224.

Funk, I. K. (1895). *The complete preacher: Sermons preached by some of the most prominent clergymen in this and other countries, and in the various denominations*. University of Michigan: Funk & Wagnalls.

Sally, D. (2000). I, too, sail past: Odysseus and the logic of self-control. *Kyklos, 53*, 173–200.

Stanford, W. (1954). *The Ulysses theme: A study in the adaptability of a traditional hero*. Ann Arbor, MI: University of Michigan Press.

Strotz, R. (1956). Myopia and inconsistency in dynamic utility maximization. *Review of Economic Studies, 23*(3), 165–180.

6 Precommitment is a term devised by Thomas Schelling, the Nobel Prize winning economist. Known for influencing fields from strategic bargaining to global warming, Schelling was also particularly good at dreaming up precommitment examples.

Schelling, T. (1984). *Choice and consequence: Perspectives of an errant economist*. Cambridge, MA: Harvard University Press.

Schelling, T. C. (1992). Self-command: A new discipline. In G. Loewenstein & J. Elster (Eds.), *Choice over time* (pp. 167–176). New York: Russell Sage Foundation.

7 O'Donoghue, T., & Rabin, M. (2008). Procrastination on long-term projects. *Journal of Economic Behavior & Organization, 66*, 161–175.

8 This lack of self-awareness is known as a "projection bias," whereby we project our present desires onto our future selves.

Loewenstein, G., & Angner, E. (2003). Predicting and indulging changing preferences. In R. F. Baumeister, G. Loewenstein & D. Read (Eds.), *Time and decision: Economic and psychological perspectives on intertemporal choice* (pp. 351–391). New York: Russell Sage Foundation.

9 That he burnt his ships is a myth, probably caused by a mistranslation or confusing the story with William the Conqueror. Regardless, it still makes for a good example.

Reynolds, W. (1959). The burning ships of Hernán Cortés. *Hispania, 42*(3), 317–324.

10 Ibeji, M. (2001). *1066: BBC History.* Retrieved from: http://www.bbc.co.uk/history/british/normans/1066_01.shtml

11 Working at my university office, I can't really use this nudist technique without people starting petitions and protests. However, it might confine me to my home, which is also a great technique. In his book on precommitment Thomas Schelling cites the *Times Literary Supplement* for January 22, 1982, in which George Steiner interviews the Hungarian radical Georg Lukacs: "When I first called on him, in the winter of 1957–8, in a house still pockmarked with shell bursts and grenade splinters, I stood speechless before the armada of his printed works, as it crowded the bookshelves. Lukacs seized on my puerile wonder and blazed out of his chair in a motion at once vulnerable and amused: 'You want to know how one gets work done? It's easy. House arrest, Steiner, house arrest!'"

Schelling, T. (1984). *Choice and consequence: Perspectives of an errant economist.* Cambridge, MA: Harvard University Press.

Wallace, I. (1977). Self-control techniques of famous novelists. *Journal of Applied Behavior Analysis, 10*(3), 515–525.

12 Weir, W. (January 12, 2006). Wake up! You snooze, you lose—Multiple hits on the snooze alarm may be hazardous to your sleep and motivation. *Newsday.*

13 Richtel, M. (June 14, 2008). Lost in E-mail, tech firms face self-made beast. *New York Times.*

Williams, A. (October 19, 2008). Drunk, and dangerous, at the keyboard. *New York Times.*

14 It is based on the Irvine Welsh book of the same name, but I have only seen the movie.

15 To underscore the importance of this wisdom, there are dozens of other sayings to this effect. For instance, George Eliot noted: "No man can be wise on an empty stomach"; Albert Einstein thought: "An empty stomach is not a good political adviser"; and William Cowper concluded: "No man can be a patriot on an empty

stomach." My favorite, though, is number 214 of the Ferengi Rules of Acquisition.

16 Not always though. As Maslow wrote: "We have spoken so far as if this hierarchy was a fixed order, but actually it is not nearly so rigid as we may have implied. It is true that most of the people with whom we have worked have seemed to have these basic needs in about the order that has been indicated. However, there have been a number of exceptions . . ."

Maslow, A. H. (1954). *Motivation and personality.* New York: Harper.

17 Cantor, N., & Blanton, H. (1996). Effortful pursuit of personal goals in daily life. In P. M. Gollwitzer & J. A. Bargh (Eds.), *The psychology of action: Linking cognition and motivation to behavior* (pp. 338–359). New York: Guilford Press.

Fiore, N. (1989). *The now habit: A strategic program for overcoming procrastination and enjoying guilt-free play*. New York: Penguin Putnam, Inc.

Schneider, F. W., & Green, J. E. (1977). The need for affiliation and sex as moderators of the relationship between need for achievement and academic performance. *Journal of School Psychology,* 15, 269–277.

18 Su, X. (2007). *A model of consumer inertia with applications to dynamic pricing*. Berkeley: University of California.

19 This form of precommitment is also known as counteractive control, contingency management, and side bets.

Loewenstein, G., & Angner, E. (2003). Predicting and indulging changing preferences. In R. F. Baumeister, G. Loewenstein, & D. Read (Eds.), *Time and decision: Economic and psychological perspectives on intertemporal choice* (pp. 351–391). New York: Russell Sage Foundation.

Milkman, K. L., Rogers, T., & Bazerman, M. (2008). *Highbrow films gather dust: A study of dynamic inconsistency and online DVD rentals.* Boston: Harvard Business School.

Moeller, F., Barratt, E., Dougherty, D., Schmitz, J., & Swann, A. (2001).

Psychiatric aspects of impulsivity. *American Journal of Psychiatry, 158*(11), 1783–1793.

Read, D., Loewenstein, G., & Kalyanaraman, S. (1999). Mixing virtue and vice: Combining the immediacy effect and the diversification heuristic. *Journal of Behavioral Decision Making* 12, 257–273.

Strotz, R. (1956). Myopia and inconsistency in dynamic utility maximization. *Review of Economic Studies, 23*(3), 165–180.

Trope, Y., & Fishbach, A. (2000). Counteractive self-control in overcoming temptation. *Journal of Personality and Social Psychology, 79*(4), 493–506.

20 Surowiecki, J. (Feb. 14, 2006). Bitter money and Christmas Clubs. *Forbes.*

21 Ashraf, N., Karlin, D., & Yin, W. (2008). *Female empowerment: Impact of a commitment savings product in the Philippines.* Boston: Jameel Poverty Action Lab. Retrieved from: http://www.povertyactionlab.org/papers/ashraf_karlan_yin_female_empowerment_0308.pdf

22 Retrieved from http://www.marginalrevolution.com/marginalrevolution/2008/09/markets-in-self.html

23 Here is one more example. To stop addicts from relapsing, a Denver cocaine addiction center encourages self-inflicted blackmail. Patients write an incriminating letter to the authorities, revealing their misdeeds and urging the strongest punitive response. If these patients then fail to pass a random series of drug tests, those letters are delivered.

Schelling, T. C. (1992). Self-command: A new discipline. In G. Loewenstein & J. Elster (Eds.), *Choice over time* (pp. 167–176). New York: Russell Sage Foundation.

24 Thaler, R., & Sunstein, C. (2008). *Nudge.* New Haven, CT: Yale University Press.

25 Lane Olinghouse.

26 Allen, K. (1996). Chronic nailbiting: A controlled comparison of

competing response and mild aversion treatments. *Behavior Research and Therapy, 34*(3), 269–272.

27 As Seymour originally bragged, "You are looking at a man who developed a foolproof system for fidelity." Richler, M. (1980). *Joshua then and now.* Toronto, ON: McClelland & Stewart.

28 Mischel, W., & Ayduk, O. (2004). Willpower in a cognitive-affective processing system. In I. Baumeister & K. Vohs (Eds.), *Handbook of self-regulation: Research, theory, and applications* (pp. 99–129). New York: Guilford Press.

29 See also: Caspi, A., Roberts, B., & Shiner, R. (2005). Personality development: Stability and change. *Annual Review of Psychology, 56*, 453–484.

Lee, P., Lan, W., Wang, C., & Chiu, H. (2008). Helping young children to delay gratification. *Early Childhood Education Journal, 35*(6), 557–564.

30 The average is over one violation per minute under attempts of active suppression. If you did make it to the minute mark, see if you can go an additional sixty seconds. It becomes much tougher.

Wenzlaff, R., & Wegner, D. (2000). Thought suppression. *Annual Reviews in Psychology, 51*(1), 59–91.

Wegner, D. (1994). *White bears and other unwanted thoughts: Suppression, obsession, and the psychology of mental control.* New York: The Guilford Press.

31 Of note, this is an inherent problem with any Panglossian approach that advocates you not think any negative thoughts. Such advice is doomed to fail by its very design.

32 Alternatively, the twentieth-century cultural critic Ernst Cassirer observed: "Physical reality seems to recede in proportion as man's symbolic activity advances."

Mischel, W., & Baker, N. (1975). Cognitive appraisals and transformations in delay behavior. *Journal of Personality and Social Psychology, 31*, 254–261.

33 Deacon, T. W. (1997). *The symbolic species.* New York: W. W. Norton & Company.

Gifford, A. (2002). Emotion and self-control. *Journal of Economic Behavior & Organization, 49,* 113–130.

Gifford, A. (2009). Rationality and intertemporal choice. *Journal of Bioeconomics, 11*(3), 223–248.

34 Tversky, A., & Kahneman, D. (1974). Judgment under uncertainty: Heuristics and biases. *Science, 185,* 1124–1131.

35 Kearney, A. (2006). A primer of covert sensitization. *Cognitive and Behavioral Practice, 13*(2), 167–175.

36 My example was actually rather mild compared to the ones Joseph Cautela, one of the originators of the technique, develops. Here Joseph describes its use in avoiding desserts:

> I want you to imagine you've just had your main meal and you are about to eat your dessert, which is apple pie. As you are about to reach for the fork, you get a funny feeling in the pit of your stomach. You start to feel queasy, nauseous, and sick all over. As you touch the fork, you can feel food particles inching up in your throat. You're just about to vomit. As you put the fork into the pie, the food comes up into your mouth. You try to keep your mouth closed because you are afraid that you'll spit the food out all over the place. You bring the piece of pie to your mouth. As you are about to open your mouth, you puke; you vomit all over your hands, the fork, over the pie. It goes all over the table, over other people's food. Your eyes are watering. Snot, mucus are all over your mouth and nose . . .

Cautela goes on (and on) but I think that is all the description you or I can probably stomach. I have no reason to hate apple pie and I would like to keep it that way. But it was effective, wasn't it?

Cautela, J. R. (1972). *Covert sensitization scenes: A compilation of typical scenes used in the application of covert sensitization to a variety of maladaptive behaviors*. Chestnut Hill, MA: Boston College.

37 Lohr, S. (September 22, 2009). A $1 million research bargain for Netflix, and maybe a model for others. *New York Times,* B1.

38 Of note, mindfulness meditation may be a relevant way of increasing your attention control, but this has yet to be proven. As described by Jon Kabat-Zinn, a molecular biologist who pioneered the practice in the West, "meditation means cultivating a non-judging attitude toward what comes up in the mind . . . to witness whatever comes up . . . and to recognize it without condemning it or pursuing it." Consequently, even if the impulse to pursue a temptation does arise, the decision to act upon this impulse is not automatic. If mindfulness meditation does prove helpful, however, I am still skeptical about its practical value. It can take a long time to effectively master and in the meantime you will find it really, really boring. This makes it exactly the type of practice that boredom-sensitive procrastinators are going to put off. In other words, if you have the patience to foster mindfulness, you probably don't need the added self-control in the first place.

Brown, K., Ryan, R., & Creswell, J. (2007). Mindfulness: Theoretical foundations and evidence for its salutary effects. *Psychological Inquiry, 18*(4), 211–237.

Kabat-Zinn, J. (1994). *Wherever you go there you are: Mindfulness meditation in everyday life*. New York: Hyperion.

Masicampo, E. J., & Baumeister, R. F. (2007). Relating mindfulness and self-regulatory processes. *Psychological Inquiry, 18*(4), 255–258.

39 Kavanagh, D. J., Andrade, J., & May, J. (2005). Imaginary relish and exquisite torture: The elaborated intrusion theory of desire. *Psychological Review, 112*(2), 446–467.

Smallwood, J., & Schooler, J. (2006). The restless mind. *Psychological Bulletin, 132*(6), 946–958.

40 Bargh, J. A., & Chartrand, T. L. (1999). The unbearable automaticity of being. *American Psychologist, 54*(7), 462–479.

Bargh, J. A., & Ferguson, M. J. (2000). Beyond behaviorism: On the automaticity of higher mental processes. *Psychological Bulletin, 126*(6), 925–945.

41 Bargh, J. (2006). What have we been priming all these years? On the development, mechanisms, and ecology of nonconscious social behavior. *European Journal of Social Psychology, 36*(2), 147–168.

Carey, B. (July 31, 2007). Who's minding the mind? *New York Times.*

42 Wansink, B. (2004). Environmental factors that increase the food intake and consumption volume of unknowing consumers. *Annual Review of Nutrition, 24,* 455–479.

43 Childress, A., Hole, A., Ehrman, R., Robbins, S., McLellan, A., & O'Brien, C. (1993). Cue reactivity and cue reactivity interventions in drug dependence. In L. S. Onken, J. D. Blaine & J. J. Boren (Eds.), *Behavioral treatments for drug abuse and dependence* (pp. 73–96). Rockville, MD: National Institute on Drug Abuse.

44 Lustig, C., Hasher, L., & Tonev, S. T. (2001). Inhibitory control over the present and the past. *European Journal of Cognitive Psychology, 13*(1), 107–122.

45 Tullier, M. (2000). *The complete idiot's guide to overcoming procrastination.* Indianapolis, IN: Alpha Books.

46 Especially see the work of psychologist Fuschia Sirios, whose work on household safety behaviors emphasizes reducing clutter, such as putting away "hazardous tools after they are used" or keeping "stairs and walkways at home free of clutter and other tripping hazards."

Sirois, F. M. (2007). "I'll look after my health, later": A replication and extension of the procrastination-health model with community-dwelling adults. *Personality and Individual Differences, 43*(1), 15–26.

47 Lay, C. H., & Schouwenburg, H. C. (1993). Trait procrastination, time management, and academic behavior. *Journal of Social Behavior & Personality, 8*(4), 647–662.

Neck, C., & Houghton, J. (2006). Two decades of self-leadership theory and research. *Journal of Managerial Psychology, 21*(4), 270–295.

48 Again, even pigeons are capable of using this type of attentional control.

Monterosso, J., & Ainslie, G. (1999). Beyond discounting: Possible experimental models of impulse control. *Psychopharmacology, 146,* 339–347.

Wenzlaff, R., & Bates, D. (2000). The relative efficacy of concentration and suppression strategies of mental control. *Personality and Social Psychology Bulletin, 26*(10), 1200.

49 There are lots of do-it-yourself kits that provide precisely this, like the *Contr014, Kill A Watt, Wattson Energy Meter* or the *Owl* (aka, *The Electrisave*); they should pay for themselves within months. Also, the hypermiler car subculture is an early adopter of this insight. With an arsenal of tricks, a few not for the faint of heart, like coasting in the draft of an eighteen-wheeler or the "death turn," they eke out incredible gas mileage just by the way they drive. But what hypermilers rave about most is a mini-computer called the *Scan Gauge,* which plugs into any car built after 1995. Velcroed prominently on your dashboard, it provides instantaneous feedback on a choice of critical outcomes like cost per mile and cost per trip, not just miles per gallon (though that is a nice start). Suddenly, driving cost-consciously and environmentally becomes upfront and second nature. Once the abstract notion of reduced gas consumption, which appeals to our prefrontal cortex, becomes more immediate, tangible, and vivid, so that it appeals to our limbic system, we will freely use less gas. For example, I have seen my thrifty mother-in-law drive for thirty minutes to a fabric store just to return one *extremely* low-cost item. Once you calculate in gas costs, the round-trip cost her money, but still she drove. Travel costs are vaguely known, while that purchase was right there in her hands. If she drove a different kind of car, one which calculated her travel costs automatically on the dashboard, I doubt she would have made the journey. This type of technology should increase our mileage by 25 percent just by

reducing idling, speeding, and unnecessary acceleration. If we could tie in the automatic "tire pressure monitoring system," indicating how much your underinflated tires are costing you, driving efficiency could increase over 3 percent. Incorporating an "air filter monitoring system" and gas mileage potentially jumps another 10 percent. Given that cars produce the bulk of greenhouse gases, this implementation alone could easily meet the targeted reductions for the Kyoto Protocol, the international environmental treaty.

Gaffney, D. (January/February 2007). This guy can get 59 MPG in a plain old Accord. Beat that, punk. *Mother Jones.*

Grunwald, M. (August, 2008). The tire-gauge solution: No joke. *Time.*

Jones, T. Y. (June, 2008). Hypermilers: Breaking the 100-MPG barrier. *Edmunds Inside Line.*

50 Thompson, C. (2007). Clive Thompson thinks: Desktop orb could reform energy hogs. *Wired, 15.08.*

51 Lohr, S. (January 10, 2008). Digital tools help users save energy, study finds. *New York Times.*

Minosi, A., Martinola, A., Mankan, S., Balzarini, F., Kostadinov, A., & Prevostini, A. (2003). *Intelligent, low-power and low-cost measurement system for energy consumption.* Paper presented at the International Symposium on Virtual Environments, Human-Computer Interfaces, and Measurement Systems, Lugano, Switzerland

52 Aarts, H., Dijksterhuis, A., & Dik, G. (2008). Goal contagion: Inferring goals from others' actions—and what it leads to. In J. Y. Shah & W. L. Gardner (Eds.), *Handbook of motivation* (pp. 265–280). New York: Guilford Press.

Gollwitzer, P., & Bargh, J. (2005). Automaticity in goal pursuit. In A. J. Elliot & C. S. Dweck (Eds.), *Handbook of competence and motivation* (pp. 624–646). New York: Guilford Press.

53 Lopez, F., & Wambach, C. (1982). Effects of paradoxical and self-control directives in counseling. *Journal of Counseling Psychology,* 29(2), 115–124.

Mulry, G., Fleming, R., & Gottschalk, A. C. (1994). Psychological reactance and brief treatment of academic procrastination. *Journal of College Student Psychotherapy, 9*(1), 41–56.

Ziesat, H. A., Rosenthal, T. L., & White, G. M. (1978). Behavioral self-control in treating procrastination of studying. *Psychological Reports, 42*, 59–69.

54 Economists actually refer to a version of stimulus cuing as mental accounting, which deals with how easily we categorize the world into discrete domains. This tendency also helps to explain the success of Christmas Clubs.

Thaler, R. (1999). Mental accounting matters. *Journal of Behavioral Decision Making, 12*, 183–206.

Surowiecki, J. (February 14, 2006). Bitter money and Christmas clubs. *Forbes.*

55 Ashforth, B. E., Kreiner, G. E., & Fugate, M. (2000). All in a day's work: Boundaries and micro role transitions. *The Academy of Management Review, 25*(3), 472–491.

56 Locke, E., & Latham, G. (2002). Building a practically useful theory of goal setting and task motivation: A 35-year odyssey. *American Psychologist, 57*(9), 705–717.

57 For example, as the management training group *RapidBi* documents on their website (http://www.rapidbi.com/created/WriteSMARTobjectives .html), the S.M.A.R.T. acronym has dozens of variations. However, people already invariably add a time frame when giving examples of specific goals. For example, *RapidBi* suggests that people should indicate "When do I want this to be completed?" when creating specific goals. Similarly, a typical definition for attainable goals is that they be "realistic." Take a look at almost any book or example on the topic.

Tayntor, C. B. (2001). Incorporating six sigma concepts into systems analysis. In P. Tinnirello (Ed.), *New directions in project management* (pp. 161–172). Boca Raton, FL: CRC Press LLC.

http://www.topachievement.com/smart.html

58 Prendergast, C. (1999). The provision of incentives in firms. *Journal of Economic Literature, 37,* 7–63.

Schlinger, H. D., Derenne, A., & Baron, A. (2008). What 50 years of research tell us about pausing under ratio schedules of reinforcement. *The Behavior Analyst, 31,* 39–40.

59 Hall, P. A., & Fong, G. T. (2003). The effects of a brief time perspective intervention for increasing physical activity among young adults. *Psychology and Health, 18*(6), 685–706.

Miller, R. B., & Brickman, S. J. (2004). A model of future-oriented motivation and self-regulation. *Educational Psychology Review, 16*(1), 9–33.

60 Engber, D. (May 15, 2008). *The unfinished stories: All the stuff we never got around to including in the special issue.* Retrieved from http://www.slate.com/id/2191420/

61 Amabile, T. (2001). Beyond talent: John Irving and the passionate craft of creativity. *American Psychologist, 56*(4), 333–336.

Wallace, I. (1977). Self-control techniques of famous novelists. *Journal of Applied Behavior Analysis, 10*(3), 515–525.

62 http://www.rescuetime.com/dashboard; http://manictime.com/

63 Ouelette, J. A., & Wood, W. (1998). Habit and intention in everyday life: The multiple processes by which past behavior predicts future behavior. *Psychological Bulletin, 124*(1), 54–74.

64 Baumeister, R. F., Muraven, M. & Tice, D. M. (2000). Ego depletion: A resource model of volition, self-regulation, and controlled processing. *Social Cognition, 18(*2), 130–150.

65 Diefendorff, J. M., Richard, E. M., & Gosserand, R. H. (2006). Examination of situational and attitudinal moderators of the hesitation and performance relation. *Personnel Psychology, 59,* 365–393.

Gollwitzer, P. M. (1996). The volitional benefits from planning. In P. M. Gollwitzer & J. A. Bargh (Eds.), *The psychology of action: Linking cognition and motivation to behavior* (pp. 287–312). New York: Guilford Press.

Silver, M. (1974). Procrastination. *Centerpoint, 1*(1), 49–54.

66 Diefendorff, J. M., Richard, E. M., & Gosserand, R. H. (2006). Examination of situational and attitudinal moderators of the hesitation and performance relation. *Personnel Psychology, 59*, 365–393.

67 McCrea, S., Liberman, N., Trope, Y., & Sherman, S. (2008). Construal level and procrastination. *Psychological Science, 19*(12), 1308–1314.

Wood, W., & Neal, D. T. (2007). A new look at habits and the habit-goal interface. *Psychological Review 114*(4), 843–863.

68 Psychologist Peter Gollwitzer calls this process *action planning* and the resulting plans *implementation intentions*. Gallo, I. S., & Gollwitzer, P. M. (2007). Implementation intentions: A look back at fifteen years of progress. *Psicothema, 19*(1), 37–42.

Gollwitzer, P., & Sheeran, P. (2006). Implementation intentions and goal achievement: A meta-analysis of effects and processes. *Advances in Experimental Social Psychology, 38*, 69–119.

Gollwitzer, P. M. (1999). Implementation intentions: Strong effects of simple plans. *American Psychologist, 54*(7), 493–503.

Owens, S., Bowman, C., & Dill, C. (2008). Overcoming procrastination: The effect of implementation intentions. *Journal of Applied Social Psychology, 38*(2), 366–384.

69 Oaten, M., & Cheng, K. (2006). Improved self-control: The benefits of a regular program of academic study. *Basic & Applied Social Psychology, 28*(1), 1–16.

Oaten, M., & Cheng, K. (2007). Improvements in self-control from financial monitoring. *Journal of Economic Psychology, 28*(4), 487–501.

And more than a few proverbs:

> "I say that habit's but a long practice, friend, and this becomes men's nature in the end."—Aristotle
>
> "Habit, if not resisted, soon becomes necessity."—St. Augustine
>
> "The chains of habit are generally too small to be felt until they are too strong to be broken."—Samuel Johnson
>
> "Habit is a cable; we weave a thread each day, and at last we cannot break it."—Horace Mann

"Man becomes a slave to his constantly repeated acts. What he at first chooses, at last compels."—Orison Swett Marden

"Habits are at first cobwebs, then cables."—Chinese Proverb

70 Wood, W., Tam, L., & Witt, M. (2005). Changing circumstances, disrupting habits. *Journal of Personality and Social Psychology, 88*(6), 918–933.

71 Grant, A. (2003). The impact of life coaching on goal attainment metacognition and mental health. *Social Behavior and Personality, 31*(3), 253–263.

72 Matlin, E. (2004). *The procrastinator's guide to wills and estate planning*. New York: Penguin Group.

CHAPTER TEN

1 Frincke, J. (2008). *Job satisfaction.* Alexandria, VA: Society for Human Resource Management.

Kaiser, R., Hogan, R., & Craig, S. (2008). Leadership and the fate of organizations. *American Psychologist, 63*(2), 96–110.

Sousa-Poza, A., & Sousa-Poza, A. A. (2000). Well-being at work: A cross-national analysis of the levels and determinants of job satisfaction. *Journal of Socio-Economics, 29*(6), 517–538.

2 Bass, B. M. (1998). *Transformational leadership: Industry, military, and educational impact.* Mahwah, NJ: Erlbaum.

Eagly, A., Johannesen-Schmidt, M., & van Engen, M. (2003). Transformational, transactional, and laissez-faire leadership styles: A meta-analysis comparing women and men. *Psychological Bulletin, 129*(4), 569–591.

Yukl, G. (2006). *Leadership in organizations* (6th ed.). Upper Saddle River, NJ: Prentice Hall.

3 Baltes, B., Briggs, T., Huff, J., Wright, J., & Neuman, G. (1999). Flexible and compressed workweek schedules: A meta-analysis of their effects on work-related criteria. *Journal of Applied Psychology, 84*(4), 496–513.

4 Tom was truly exceptional. In survey after survey and study after study, about three quarters of employees report that the *worst* aspect of their job is their immediate supervisor, and about two thirds of supervisors would be considered incompetent by any objective standards.

Hogan, R., & Kaiser, R. (2005). What we know about leadership. *Review of General Psychology*, 9(2), 169–180.

5 Milgram, N. A. (1991). Procrastination. In R. Dulbecco (Ed.), *Encyclopedia of human biology* (Vol. 6, pp. 149–155). New York: Academic Press.

6 Ainslie, G. (2001). *Breakdown of will.* Cambridge University Press.

Ryan, R. M., & Deci, E. L. (2006). Self-regulation and the problem of human autonomy: Does psychology need choice, self-determination, and will? *Journal of Personality & Social Psychology*, 74(6), 1557–1586.

Vohs, K. D., & Baumeister, R. F. (2007). Can satisfaction reinforce wanting? In J. Y. Shah & W. L. Gardner (Eds.), *Handbook of motivation science* (pp. 373–389). New York: Guilford Press.

7 Kivetz, R., & Keinan, A. (2006). Repenting hyperopia: An analysis of self-control regrets. *Journal of Consumer Research, 33*, 273–282.

8 Tangney, J., Baumeister, R., & Boone, A. (2004). High self-control predicts good adjustment, less pathology, better grades, and interpersonal success. *Journal of Personality*, 72(2), 271–324.

POSTSCRIPT

1 Carver, C. S. (2005). Impulse and constraint: Perspectives from personality psychology, convergence with theory in other areas, and potential for integration. *Personality and Social Psychology Review*, 9(4), 312–333.

Cervone, D., Shadel, W. G., Smith, R. E., & Fiori, M. (2006). Self-regulation: Reminders and suggestions from personality science. *Applied Psychology: An International Review*, 55(3), 333–385.

Mesoudi, A., Whiten, A., & Laland, K. (2006). Towards a unified science of cultural evolution. *Behavioral and Brain Sciences, 29*(4), 329–347.

Tooby, J., & Cosmides, L. (2007). Evolutionary psychology, ecological rationality, and the unification of the behavioral sciences. *Behavioral and Brain Sciences, 30*(1), 42–43.

2 Green, C. D. (1992). Is unified positivism the answer to psychology's disunity? *American Psychologist, 47,* 1057–1058.

Staats, A. W. (1999). Unifying psychology requires new infrastructure, theory, method, and a research agenda. *Review of General Psychology, 3*(1), 3–13.

Stanovich, K. E. (2007). The psychology of decision making in a unified behavioral science. *Behavioral and Brain Sciences, 30*(1), 41–42.

3 It is why one of my key articles is titled *Integrating Theories of Motivation.*

Steel, P. & König, C. J. (2006). Integrating theories of motivation. *Academy of Management Review, 31,* 889–913.

4 Wilson, E. (1998). *Consilience: The unity of knowledge.* New York: Knopf.

5 Gintis, H. (2004). Towards the unity of the human behavioral sciences. *Politics, Philosophy & Economics, 3*(1), 37–57.

6 Akerlof, G. A. (1991). Procrastination and obedience. *American Economic Review, 81*(2), 1–19.

Glimcher, P., & Rustichini, A. (2004). Neuroeconomics: The consilience of brain and decision. *Science, 306,* 447–452.

7 Kubey, R., & Csikszentmihalyi, M. (2002). Television addiction is no mere metaphor. *Scientific American, 286*(2), 62–68.

Young, K. (1998). Internet addiction: The emergence of a new clinical disorder. *Cyberpsychology and Behavior, 1,* 237–244.

8 Hancox, R., & Poulton, R. (2006). Watching television is associated with childhood obesity: but is it clinically important? *International Journal of Obesity, 30,* 171–175.

Vandewater, E., Bickham, D., & Lee, J. (2006). Time well spent? Relating television use to children's free-time activities. *Pediatrics, 117*(2), 181–191.

9 Hall, L., Johansson, P., & Léon, D. d. (2002). *The future of self-control: Distributed motivation and computer-mediated extrospection.* Lund: Lund University.

WORKBOOK: A STEP-BY-STEP GUIDE TO GETTING STUFF DONE

1 Calorie Control Council (2007), "Majority of Americans think about dieting year-round," retrieved from www.caloriecontrol.org/pr_08092007-b.html; Consumer Reports (May 2007), "Dieters are sure they'll win at losing." Retrieved from: www.consumerreports.org/health/healthy-living/diet-nutrition/diets-dieting/dieters-are-optimistic-about-weight-loss-survey-shows/overview/dietpoll.htm.

2 Cooper, G. (January 9, 2008), "Third of all adults constantly on a diet," Reuters; Herbalife (March 22, 2004), "Optimism about future weight loss surges while the worldwide waistline widens, according to new survey," *Business Wire*; Rosenthal, E. (May 4, 2005), "Even the French are fighting obesity," *International Herald Tribune*; Rosenthal, E. (September 24, 2008), "Fast food hits Mediterranean; a diet succumbs," *New York Times.*

3 "India becoming land of obese" (November 25, 2008), *The Hindu*; Rosin, O. (2008), "The economic causes of obesity: A survey," *Journal of Economic Surveys, 22*(4), 1–31.

4 James, P. (2004), "Obesity: The worldwide epidemic," *Clinics in Dermatology,* 22(4), 276–80; WHO Europe (2005), "The challenge of obesity in the WHO European Region." Fact sheet EURO, 13.

5 Critchfield, T. and Kollins, S. (2001), "Temporal discounting: Basic research and the analysis of socially important behavior," *Journal of Applied Behavior Analysis, 34*(1), 101–122; Dodd, M. (2008), "Obesity and time-inconsistent preferences," *Obesity Research & Clinical Practice,* 2(2), 83–89; Finkelstein, E., Ruhm, C., & Kosa, K.

(2005), "Economic causes and consequences of obesity," *Annual Review of Public Health, 26*(1), 239–257; Stutzer, A. (2006), *When temptation overwhelms will-power: Obesity and happiness*, working paper, University of Basel.

6 Tullier, L. (2000), *The Complete Idiot's Guide to Overcoming Procrastination* (Indianapolis, IN: Alpha Books).

7 Campos, P., Saguy, A., Ernsberger, P., Oliver, E., & Gaesser, G. (2006), "The epidemiology of overweight and obesity: public health crisis or moral panic?" *International Journal of Epidemiology, 35*(1), 55–60; Lee, D., Sui, X., & Blair, S. N. (2009), "Does physical activity ameliorate the health hazards of obesity?" *British Journal of Sports Medicine, 43*(1), 49-51.

8 Carr, N. (2010), *The Shallows: What the Internet Is Doing to Our Brains* (New York: W.W. Norton & Company).

9 D'Abate, C., & Eddy, E. (2007), "Engaging in personal business on the job: Extending the presenteeism construct," *Human Resource Development Quarterly, 18*(3), 361–83; Johnson, P. R., & Indvik, J. (2003), "The organizational benefits of reducing cyberslacking in the workplace," *Proceedings of the Academy of Organizational Culture, Communications and Conflict,* 7(2), 53–59; Lavoie, J. A. A., & Pychyl, T. A. (2001), "Cyberslacking and the procrastination superhighway: A web-based survey of online procrastination, attitudes, and emotion." *Social Science Computer Review, 19*(4), 431–444; Malachowski, D. (2005), "Wasted time at work costing companies billions," Salary.com.

10 Byrne, A., Blake, D., Cairns, A., & Dowd, K. (2006), "There's no time like the present: the cost of delaying retirement saving," *Financial Services Review, 15*(3), 213–231; Steel, P. (2007), "The nature of procrastination," *Psychological Bulletin, 133*(1), 65–94.

11 "Canada's personal debt rises" (June 1, 2011), CBC News, retrieved from: www.cbc.ca/news/business/story/2011/06/01/canada-cosumer-credit.html.

12 "Indebtedness after the financial crisis," (June 24, 2010), *The Economist*, retrieved from: www.economist.com/blogs/buttonwood/2010/06/indebtedness_after_financial_crisis.

13 Often studied under the term bibliotherapy; see Bergsma, A. (2008), "Do self-help books help?" *Journal of Happiness Studies*, 9, 341–360.

INDEX

Boldface numerals denote graphs and charts.

PIERS STEEL, PhD, is the world's leading researcher and speaker on the science of motivation and procrastination. He studied and taught at the business and psychology schools of the University of Minnesota before moving to the University of Calgary's Haskayne School of Business, where he is a professor of human resources and organizational dynamics. He has been studying procrastination and its effects for more than ten years, and has spent the decades before that practicing it. Dr. Steel's award-winning research has appeared in magazines ranging from *Psychology Today* and *New Scientist* to *Good Housekeeping* and *Profit*. His work has been reported in the *Los Angeles Times*, the *Wall Street Journal*, the *New York Times*, and *USA Today*. Winner of the Killam Emerging Research Leader Award, he lives in Calgary, Alberta, with his wife and two sons.